endorsed by
edexcel :::

D0230773

A Level
Mathematics
for Edexcel

Statistics

S1

LEARNING CENTRE

| CLASS NO: | 519·5 NIC |
| ACC NO: | 072836 |

James Nicholson

HARROW COLLEGE
HH Learning Centre
Lowlands Road, Harrow
Middx HA1 3AQ
020 8909 6520

OXFORD
UNIVERS

HARROW COLLEGE

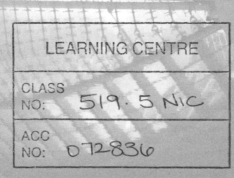

OXFORD
UNIVERSITY PRESS

Great Clarendon Street, Oxford OX2 6DP

Oxford University Press is a department of the University of Oxford.
It furthers the University's objective of excellence in research, scholarship,
and education by publishing worldwide in

Oxford   New York

Auckland   Cape Town   Dar es Salaam   Hong Kong   Karachi
Kuala Lumpur   Madrid   Melbourne   Mexico City   Nairobi
New Delhi   Shanghai   Taipei   Toronto

With offices in

Argentina   Austria   Brazil   Chile   Czech Republic   France   Greece
Guatemala   Hungary   Italy   Japan   South Korea   Poland   Portugal
Singapore   Switzerland   Thailand   Turkey   Ukraine   Vietnam

Oxford is a registered trade mark of Oxford University Press
in the UK and in certain other countries

© Oxford University Press 2008

The moral rights of the author have been asserted

Database right Oxford University Press (maker)

First published 2008

All rights reserved. No part of this publication may be reproduced,
stored in a retrieval system, or transmitted, in any form or by any means,
without the prior permission in writing of Oxford University Press,
or as expressly permitted by law, or under terms agreed with the appropriate
reprographics rights organization. Enquiries concerning reproduction
outside the scope of the above should be sent to the Rights Department,
Oxford University Press, at the address above

You must not circulate this book in any other binding or cover
and you must impose this same condition on any acquirer

British Library Cataloguing in Publication Data

Data available

ISBN: 978 0 19 911782 6

10 9 8 7 6 5 4 3 2 1

Printed in Great Britain by Ashford Colour Press Ltd, Gosport

Paper used in the production of this book is a natural, recyclable product made from wood
grown in sustainable forests. The manufacturing process conforms to the environmental
regulations of the country of origin.

Acknowledgements

The photograph on the cover is reproduced
courtesy of Jose Fuste Raga/Image State

The publishers and authors would like to thank the following for permission to use
photographs: **p 5** iStockphoto / Damon Bell
p16 Getty/Image Source; **p18** Corbis UK Ltd / Firefly Productions; **p20** (top) Alamy/Bernhard
Classen; **p20** (lower) Alamy/Tom McCarthy; **p21** (top) Getty Images / Christopher Furlong OUP;
p21 (lower) OUP **p24** OUP; **p27** Alamy / Andrew Darrington; **p56** Alamy / Israel Images;
p88 istockphoto; **p96** OUP/ Texas Instruments; **p110** Alamy; **p140** Alamy/Dennis Hallinan;
p166 (top) Alamy/David Hoffman (lower) OUP;   **p167** (left) Nir Elias/Reuters/Corbis, (right)
Bettmann/Corbis; **p182** Alamy/Eyebyte; **p190** OUP..

The publishers would also like to thank Judy Sadler and Ian Bettison for their expert help in
compiling this book.

HARROW COLLEGE
HH Learning Centre
Lowlands Road, Harrow
Middx HA1 3AQ
020 8909 6520

About this book

Endorsed by Edexcel, this book is designed to help you achieve your best possible grade in Edexcel GCE Mathematics Statistics S1 (6683) for first examination in January 2009. This unit can contribute toward an AS level or an A level in Mathematics.

Each chapter starts with a list of objectives and a 'Before you start' section to check that you are fully prepared. Chapters are structured into manageable sections, and there are certain features to look out for within each section:

Key points are highlighted in a blue panel.

Key words are highlighted in bold blue type.

Worked examples demonstrate the key skills and techniques you need to develop.

EXAMPLE 2

X is a random variable with probability distribution given by

x	-2	-1	0	1	2
$P(X=x)$	0.1	0.1	0.4	a	0.1

Find the value of a.

$\Sigma p_i = 1$
$1 - (0.1 + 0.1 + 0.4 + 0.1) = 0.3$
$\therefore a = 0.3$

Helpful hints are included as blue margin notes and sometimes as blue type within the main text.

Each section includes an exercise with progressive questions, starting with basic practice and developing in difficulty. Some exercises also includes 'stretch and challenge' questions marked with a stretch symbol ▌.

At the end of each chapter there is a 'Review' section which includes exam style questions as well as past exam paper questions.

The final page of each chapter gives a summary of the key points, fully cross-referenced to aid revision. Also, a 'Links' feature provides an engaging insight into how the statistics you are studying is relevant to real life.

There are also two Revision Exercises within the book which contain questions spanning a range of topics to give you plenty of realistic exam practice.

At the end of the book you will find full solutions, a key word glossary, a list of essential formulae, statistical tables and an index.

Contents

Background statistics

This chapter will show you how to
- recap types of data
- recap basic statistical calculations
- recap graphical representations.

Before you start

You should know how to:

1 Calculate the mean, median, range and mode of a simple set of data.

e.g. 7, 5, 3, 4, 7, 6, 7, 5, 9, 5, 5

The median is the middle value with the data in order.
3, 4, 5, 5, 5, **5**, 6, 7, 7, 7, 9
The median is 5.

The range = highest value – lowest value
3, 4, 5, 5, 5, 5, 6, 7, 7, 7, **9**
The range = 9 – 3
= 6

The mode is the value which appears most often.
3, 4, **5, 5, 5, 5**, 6, 7, 7, 7, 9
The mode is 5.

The mean $= \dfrac{\text{sum of the values}}{\text{number of values}}$

$= \dfrac{(7 + 5 + 3 + 4 + 7 + 6 + 7 + 5 + 9 + 5 + 5)}{11}$

$= \dfrac{63}{11}$

$= 5.7$ (to 1 d.p.)

Check in

1 **a** For the data 3, 5, 2, 0, 2, 1, 3, 2, 0 calculate
   i the mean
   ii the median
   iii the range
   iv the mode.

b For the data 7, 8, 6, 6, 6, 8 calculate
   i the mean
   ii the median
   iii the range
   iv the mode.

Types of data

You can describe this blanket by its properties. For example:

- it is blue
- it is rectangular
- it is made of wool

- it is 1.2 m by 90 cm
- it weighs 1.3 kg
- it is 5 mm thick

These are **qualitative data** – non-numerical.

These are **quantitative data** – numerical.

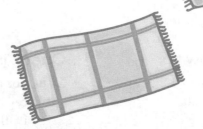

There are two types of quantitative data.

The number of blankets can only be one of a list of values (0, 1, 2, ...) This is an example of **discrete data**.

The length of the blanket is given as 1.2 m, but if you measured it more exactly, you might find it is 118.73 cm (to 2 d.p.) This is **continuous data** – the actual value can never be stated exactly – it will always be reported as a rounded value.

- Data can either be quantitative (numerical) or qualitative (non-numerical).
- Quantitative data can either be discrete (limited to particular values) or continuous (any value within a range).

EXAMPLE 1

What data about the radio are given in the advertisement? For each piece of the data, state whether it is qualitative, discrete or continuous.

Power output – continuous
Wavebands – qualitative
Number of batteries needed – discrete
Price – discrete
Colour – qualitative
Length, width, height – continuous

Puretone
DAB/FM radio
£ 89.99

Power output	2 × 2 W
Wavebands	DAB/FM
Dimensions	25 cm × 20 cm × 5 cm
Colour	Available in black or silver
Mains/battery	Mains, or 4 × AA batteries

For a large number of data observations, a list is not easy to interpret. A **frequency table** of values can be used to summarise a large list. Sometimes it is helpful to use groups of values so the data are more easily understood.

EXAMPLE 2

A group of 45 women goes on a skiing holiday.
Summarise the data below in frequency tables.

a The number of children they have:

1	0	2	1	2	0	0	3	1	1	2	1	0	0	1
2	1	1	0	1	1	1	0	2	5	1	2	1	0	1
2	1	3	1	1	0	0	1	2	1	1	3	1	0	0

b Their ages [use class intervals of 15–19, 20–24, …, 45–49]

22	26	26	45	26	27	17	29	35	38	25	23	17	38	48
26	28	35	32	19	28	35	17	29	36	32	34	27	19	25
28	23	35	24	29	37	20	30	44	25	19	32	22	22	34

a

Number of children	Frequency
0	12
1	21
2	8
3	3
4	0
5	1

b

Age	Frequency
15–19	6
20–24	7
25–29	15
30–34	6
35–39	8
40–44	1
45–49	2

15 women are aged 25–29

Exercise 1.1

1 Chris Hoy won the men's Olympic
1 km cycle race in Athens in 2004
with a time of 60.711 seconds.

For each of the following identify
the type of data:

a the time Hoy took to complete
the race

b the colour of the helmet he wore

c the number of spokes in
the front wheel of his bicycle

d the country he represented in
the Olympics

e the number of spectators in
the stadium during his race.

2 The number of days each student in a class
was late during a week is shown below.
Summarise these data in a frequency table.

0	1	0	0	1	3	0	0	0	0
0	5	1	0	1	0	0	0	0	2
0	1	1	0	0	0	1	2	0	0

3 The depth of water (in metres) at high tide
was recorded at London Bridge for a period
of 15 days in March 2006. The data are listed
below. Summarise these data in a grouped
frequency table. Use appropriate class intervals.

7.2	7.5	7.3	7.4	7.3	7.3	7.3	7.3	7.1	7.1
6.8	6.9	6.3	6.4	5.9	5.9	5.4	5.4	5.1	5.5
5.4	5.9	5.9	6.3	6.3	6.7	6.6	6.8	6.6	6.7

One way to present data which keeps the detail is to use a **stem and leaf diagram**.

You should be familiar with these from GCSE.

The data on depths at high tide in Exercise 1.1, Question 3 look like this when put into a stem and leaf diagram:

*This is a **stem**. It represents 6.0–6.9.*

```
5 | 4  4  1  4  9  9  5  9  9
6 | 3  4  3  3  8  9  7  6  8  6  7
7 | 2  3  4  3  3  3  3  1  1  5
```

*This is a **leaf**. It represents 5.9 metres.*

You can split each stem into two, to give more shape to the distribution:

This represents 6.0–6.4.

```
5 | 4  4  1  4
5 | 9  9  5  9  9
6 | 3  4  3  3
6 | 8  9  7  6  8  6  7
7 | 2  3  4  3  3  3  3  1  1
7 | 5
```

Now rewrite with the data in order, to give a completed stem-and-leaf diagram:

```
5 | 1  4  4  4                          (4)
5 | 5  9  9  9  9                       (5)
6 | 3  3  3  4                          (4)
6 | 6  6  7  7  8  8  9                 (7)
7 | 1  1  2  3  3  3  3  3  4           (9)
7 | 5                                   (1)
```

The numbers in brackets represent frequencies.

Key: 6|6 means a depth at high tide of 6.6 metres

In a stem and leaf diagram individual values of the data are preserved.

Exercise 1.2

In each question, decide whether to split each stem into two or not. Don't forget to include a key.

1 Old Faithful is a geyser in Yellowstone National Park in the USA, which is a major tourist attraction. The times (in minutes, to the nearest minute) between eruptions are recorded:

71	57	80	75	55	60	86	77	56	81
50	89	54	90	73	60	83	65	82	84
54	85	58	79	57	88	68	76	78	74
85	75	65	76	58	91	50	87	48	93

Put these data into a stem and leaf diagram.

2 The total head lengths of a sample of treecreeper birds are recorded in mm, and the results are shown below.

| 293 | 297 | 301 | 315 | 303 | 324 | 344 |
| 298 | 322 | 315 | 319 | 326 | 337 | 311 |

Represent these data in a stem and leaf diagram.

3 A group of Year 12 students recorded their pulse rates when they were in a relaxed state.

67 62 58 75 62 79 63 55 71 64 69 57

64 59 73 62 69 63 68 55 83 59 62 67

80 71 67 63 65 59 70 62 65 68 54 67

Show these data in a stem and leaf diagram.

4 The length of time, in minutes, that customers spend in a coffee shop is recorded. The results are shown below.

17, 15, 9, 31, 33, 41, 8, 14, 13, 22, 27, 43, 32, 14

Draw a stem and leaf diagram to represent these data.

5 An airport bus service runs from the city centre, and from the airport, every 10 minutes during the day. The number of passengers on a random sample of journeys from the airport is shown below.

11, 3, 14, 34, 1, 6, 12, 15, 8, 4, 28, 7, 3, 0, 8, 12

Draw a stem and leaf diagram to represent these data.

Histograms can be used to show continuous data or grouped discrete data.

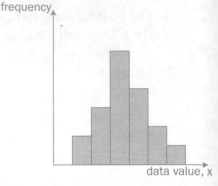

frequency

data value, x

> In a histogram
> o the area of each bar is proportional to the frequency of the class interval
> o the x-axis is a continuous linear scale
> o when the class intervals are equal, the height of each bar will also be proportional to the frequency.
> o each bar starts at the lower class boundary (lcb) and ends at the upper class boundary (ucb).

Here are the Old Faithful data from page 5.

71	57	80	75	55	60	86	77	56	81
50	89	54	90	73	60	83	65	82	84
54	85	58	79	57	88	68	76	78	74
85	75	65	76	58	91	50	87	48	93

This interval starts at 59.5 and ends at 64.5. It includes 60, 61, 62, 63 and 64.

You can plot these data on a histogram:

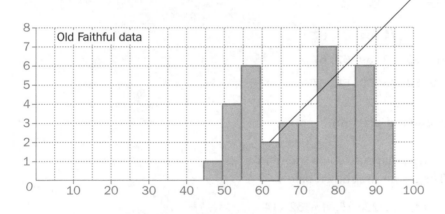

Old Faithful data

Exercise 1.3

1 The depths of water at high tide at Margate during March 2006 are summarised in the following table.

Depth of water (in metres to 1 d.p.)	Frequency
3.5–3.9	13
4.0–4.4	19
4.5–4.9	24
5.0–5.4	5

The intervals will be:
3.45–3.95
3.95–4.45
4.45–4.95
4.95–5.45

Draw a histogram to represent these data.

2 Draw a histogram for the ages in Example 2 on page 3.

Age	Frequency
15–19	6
20–24	7
25–29	15
30–34	6
35–39	8
40–44	1
45–49	2

Be careful with class boundaries for age. For example, you are 16 until your 17th birthday.

3 A large group of teenagers started a regular exercise programme. Their normal pulse rates were measured, and the data are shown in the table.

Normal pulse rate	Frequency
60–69	5
70–79	8
80–89	22
90–99	29
100–109	13
110–119	5

Draw a histogram to show these data.

4 After a series of short warm-up exercises, the group of teenagers in question 3 had their pulse rates measured again.

Pulse rate after warm-up exercises	Frequency
80–89	9
90–99	12
100–109	25
110–119	17
120–129	9
130–139	7
140–149	3

Draw a histogram to show these data.

S1

EXAMPLE 1

Remember, for a set of data values:

o the mean is the sum of all the values divided by the number of values
o the median is the middle value when the values are arranged in order
o the mode is the most commonly occurring value.

A small class was given a short mental arithmetic test.
Their scores were
7, 6, 7, 5, 6, 8, 5, 7, 8, 9
For this class calculate:

a the mean   b the median   c the mode

a The sum of the 10 scores is 68, so the mean is $\frac{68}{10} = 6.8$

b In order the scores are 5, 5, 6, 6, 7, 7, 7, 8, 8, 9.

The median is the middle of 5th and 6th scores, which are both 7, so the median is 7.

c The score 7 occurs more than any other score, so the mode is 7.

Data in a frequency table

Example 2 on page 3 shows data about a group of 45 women who went on a skiing holiday. The number of children they have is shown in the table.

Number of children	Frequency f
0	12
1	21
2	8
3	3
4	0
5	1

This is the greatest frequency, so the mode is one child.

▌Note that the mode is not 21.

There are 45 values, so the median is the 23rd value in order.
Make a running total of the frequencies:
12 women with no children, 33 women with none or one child.
The median number of children = 1.

S1

To calculate the mean, construct a table of values, with a column showing the 'value × frequency' or xf.

Number of children x	Frequency f	xf
0	12	0
1	21	21
2	8	16
3	3	9
4	0	0
5	1	5
	$\Sigma f = 45$	$\Sigma xf = 51$

Eight women with two children each makes 16 children in total.

Write Σf for the 'sum of the frequencies'.

The mean number of children is $\frac{51}{45} = 1.1$ (to 1 d.p.)

Data in a grouped frequency table

In the same example, the ages of the 45 women are

Age	Frequency
15–19	6
20–24	7
25–29	15
30–34	6
35–39	8
40–44	1
45–49	2

The modal class is 25–29 years, as this class has the highest frequency.

▌The mode is not 15.

S1

There are 45 women, so the median is the 23rd age.
Add up the frequency column:
6 women less than 20 years old
$6 + 7 = 13$ women less than 25 years old
$13 + 15 = 28$ less than 30 years old.
The median age lies in the 25–29 age group.

To calculate the mean you add up all the values but, because the frequencies are grouped, you have to use the mid-interval value and make an **estimate**.

Age	Frequency, f	Mid-interval, m	mf
15–19	6	17.5	105
20–24	7	22.5	157.5
25–29	15	27.5	412.5
30–34	6	32.5	195
35–39	8	37.5	300
40–44	1	42.5	42.5
45–49	2	47.5	95
	$\Sigma f = 45$		$\Sigma mf = 1307.5$

Remember you are 19 until your 20th birthday.

You will meet a way of simplifying the arithmetic on page 42.

So the mean age is $\frac{1307.5}{45} = 29.06$ or about 29 years old.

Exercise 1.4

1 A golfer keeps a record of his scores in club competitions during 2007. They are 75, 77, 77, 74, 79, 76, 84, 76, 75, 77. Calculate

  a the mode

  b the median

  c the mean of his scores.

2 A vet keeps a record of the number of kittens in litters produced by cats in his practice.

Number in litter	Frequency
1	2
2	4
3	7
4	11
5	8
6	4
7	2
8	1

Calculate

  a the mode

  b the median

  c the mean size of litter.

3 In Exercise 1.3, question 3 and question 4, the pulse rates for a group of teenagers were given under two conditions.
For each condition, calculate an estimate of the mean pulse rate for the group.
Use your results to compare the distributions.

Normal pulse rate	Frequency	Pulse rate after warm-up exercises	Frequency
60–69	5	80–89	9
70–79	8	90–99	12
80–89	22	100–109	25
90–99	29	110–119	17
100–109	13	120–129	9
110–119	5	130–139	7
		140–149	3

4 The table shows information about the heights of a number of plants of a particular species.

Length (cm)	20–50	50–60	60–65	65–70	70–80	80–90	90–110
Number of plants	27	18	16	15	22	14	14

The class boundaries are 20, 50, 60, …

Calculate an estimate of the mean height of these plants.

5 A question on a survey asked the children how long their journeys to school had taken on a particular day. A summary of the results is shown below.

Time (minutes, correct to the nearest minute)	Number of children
1–15	115
16–25	46
26–35	36
36–55	22
56–80	14

Calculate an estimate of the mean time the children took to come to school that day.

6 The table shows information about the salaries paid to employees in a company.

Salary	Frequency
£0 $< x \leqslant$ £10 000	7
£10 000 $< x \leqslant$ £15 000	82
£15 000 $< x \leqslant$ £20 000	45
£20 000 $< x \leqslant$ £25 000	24
£25 000 $< x \leqslant$ £30 000	13
£30 000 $< x \leqslant$ £50 000	4

Remember:
$<$ means less than
$\leqslant$ means less than or equal to

Calculate an estimate of the mean salary of employees in this company.

S1

A useful way of organising data is to divide then into quarters.
The dividing values are known as quartiles.

These dots represent 19 values:

○ ○ ○ ○ ○ ○ ○ ○ ○ ○ ○ ○ ○ ○ ○ ○ ○ ○ ○

Q_1 is the lower quartile.

For a dataset with n values $(x_1, x_2, \ldots, x_n)$ calculate $\frac{1}{4}n$.

If $\frac{1}{4}n$ is an integer r then Q_1 is the midpoint of x_r and x_{r+1}.

If $\frac{1}{4}n$ lies between r and $r+1$ then Q_1 is x_{r+1}.

$\frac{1}{4} \times 19 = 4\frac{3}{4}$, so Q_1 is the 5th value.

Q_2 is the median.

To find the median, calculate $\frac{1}{2}n$.

If $\frac{1}{2}n$ is an integer r then Q_2 is the midpoint of x_r and x_{r+1}.

If $\frac{1}{2}n$ lies between r and $r+1$ then Q_2 is x_{r+1}.

$\frac{1}{2} \times 19 = 9\frac{1}{2}$, so Q_2 is the 10th value.

Q_3 is the upper quartile.

To find the upper quartile, calculate $\frac{3}{4}n$.

If $\frac{3}{4}n$ is an integer r then Q_3 is the midpoint of x_r and x_{r+1}.

If $\frac{3}{4}n$ lies between r and $r+1$ then Q_3 is x_{r+1}.

$\frac{3}{4} \times 19 = 14\frac{1}{4}$, so Q_3 is the 15th value.

The range $= Q_4 - Q_0$

The range indicates the spread of the values.

The interquartile range (IQR) $= Q_3 - Q_1$

The IQR indicates the spread of the middle 50% of values.

EXAMPLE 1

In Section 1.4 the scores on a mental arithmetic test in a small class were given.

5, 5, 6, 6, 7, 7, 7, 8, 9, 9

Calculate:

a the quartiles   b the interquartile range.

----

a $n = 10$, so $\frac{1}{4}n = 2.5$ and Q_1 is the 3rd value. $Q_1 = 6$

$\frac{1}{2}n = 5$, so take the midpoint of the 5th and 6th values

These are both 7, so the median, $Q_2 = 7$.

$\frac{3}{4}n = 7.5$ and Q_3 is the 8th value, so $Q_3 = 8$.

b IQR $= 8 - 6 = 2$

S1

You can quickly identify quartiles when data are presented in a stem and leaf plot.

For the data on high tides given in question 3 on page 3, $n = 30$, so $\frac{1}{4}n = 7.5$, $\frac{1}{2}n = 15$, $\frac{3}{4}n = 22.5$

so the quartiles are the 8th and 23rd values, and the median is the midpoint of the 15th and 16th values.

```
5 | 1  4  4  4                     (4)
5 | 5  9  9  9  9                  (5)
6 | 3  3  3  4                     (4)
6 | 6  6  7  7  8  8  9            (7)
7 | 1  1  2  3  3  3  3  3  4      (9)
7 | 5                              (1)
```

Key: 6|6 means a depth at high tide of 6.6 metres

The lower quartile is 5.9 m, the median is 6.65 m and the upper quartile is 7.2 m. The IQR = 7.2 m − 5.9 m = 1.3 m

Exercise 1.5

1 A golfer keeps a record of his scores in club competitions during 2007. They are 75, 77, 77, 74, 79, 76, 84, 76, 75, 77. Calculate the median, lower and upper quartiles for these data.

2 A vet keeps a record of the number of kittens in litters produced by cats in his practice.

Calculate the interquartile range for these data.

Number in litter	Frequency
1	2
2	4
3	7
4	11
5	8
6	4
7	2
8	1

3 Old Faithful is a geyser in Yellowstone National Park in the USA. The times (in minutes, to the nearest minute) between eruptions are recorded and displayed in a stem and leaf diagram below.

```
4 | 8                                    (1)
5 | 0  0  4  4  5  6  7  7  8  8         (10)
6 | 0  0  5  5  8                        (5)
7 | 1  3  4  5  5  6  6  7  8  9         (10)
8 | 0  1  2  3  4  5  5  6  7  8  9      (11)
9 | 0  1  3                              (3)
```

Key: 6|5 means an interval of 65 minutes between eruptions.

Find the median and quartiles of these times.

Boxplots are good at giving a quick impression of the most important features of a distribution – location, spread and shape.

You may be familiar with these from GCSE.

Min value Q_1 Q_2 Q_3 Max value

Here are two examples:

1 $Q_0 = 15$, $Q_1 = 19.7$, $Q_2 = 22.3$, $Q_3 = 25.1$, $Q_4 = 30.4$

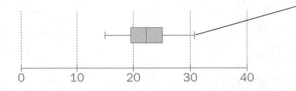

The diagram is sometimes known as a 'box and whisker plot' because the lines look like whiskers.

The distribution is fairly symmetrical, and the two outside lines are wider than the two middle box sections, showing that more values occur between 20 and 25 than elsewhere.

2 $Q_0 = 10$, $Q_1 = 18.6$, $Q_2 = 26.3$, $Q_3 = 32.1$, $Q_4 = 37.1$

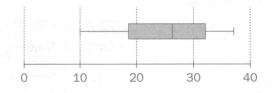

The data are more spread out at lower values than at higher values, and are not concentrated towards the centre of the distribution.

In Section 3.2 you will look at identifying unusual values, or outliers.
In Section 3.7 you will look at measures of asymmetry, or skewness.

Both of these properties can be easily seen in a boxplot.

Exercise 1.6
For questions 1 to 3, you can use the medians and quartiles you calculated for these data sets in Exercise 1.5 on page 13.

1 A golfer keeps a record of his scores in club competitions during 2007. They are 75, 77, 77, 74, 79, 76, 84, 76, 75, 77. Show these data in a boxplot.

2 A vet keeps a record of the number of kittens in litters produced by cats in his practice.

Draw a boxplot to represent these data.

Number in litter	Frequency
1	2
2	4
3	7
4	11
5	8
6	4
7	2
8	1

3 Old Faithful is a geyser in Yellowstone National Park in the USA. The times (in minutes, to the nearest minute) between eruptions are recorded and displayed in a stem and leaf diagram below.

```
4 | 8                                              (1)
5 | 0   0   4   4   5   6   7   7   8   8         (10)
6 | 0   0   5   5   8                              (5)
7 | 1   3   4   5   5   6   6   7   8   9         (10)
8 | 0   1   2   3   4   5   5   6   7   8   9     (11)
9 | 0   1   3                                      (3)
```

Key: 6|5 means an interval of 65 minutes between eruptions.

Show these data in a boxplot.

4 A group of Year 12 students recorded their pulse rates when they were in a relaxed state. The ordered data are given below.

53	54	55	55	57	58	59	59	59	62	62	62
62	62	63	63	63	64	64	65	65	67	67	67
68	68	69	69	70	71	71	73	75	79	80	83

Show these data in a boxplot.

5 The number of days each pupil was late during a week was recorded for a class, and is shown below.

```
0   1   0   0   1   3   0   0   0   0
0   5   1   0   1   0   0   0   0   2
0   1   1   0   0   0   1   2   0   0
```

Show these data in a boxplot.

Summary

You should know how to construct and use:

- ○ stem and leaf diagrams
- ○ boxplots
- ○ histograms for equal class intervals.

You should be able to calculate:

- ○ mean, median and mode from data presented
  - ○ in a list
  - ○ in a frequency table
  - ○ in a grouped frequency table
- ○ quartiles, the interquartile range and the range
  - ○ from data
  - ○ from a stem and leaf diagram
  - ○ from a boxplot.

Links

Data mining is an analytic process designed to explore data (usually large amounts of data – typically business or market related) in search of consistent patterns and/or systematic relationships between variables.

The process of data mining consists of three stages:
1) the initial exploration
2) model building or pattern identification
3) the application of the model to new data in order to generate predictions.

Mathematical models

This chapter will show you how to
- understand what a model is
- understand how mathematics is used to model real-life situations
- understand the role of probability and statistics in modelling real-world situations in which there is uncertainty or unpredictability.

You may be familiar with the data handling cycle from earlier work in statistics at school.

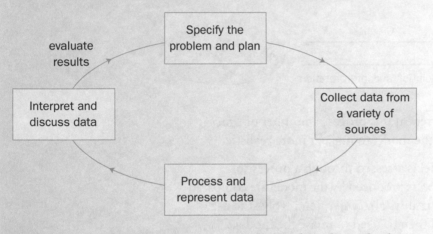

This chapter explores how statistics are used to describe and solve real-life problems involving data.

Models imitate the behaviour of real situations to some extent. However, while a radio controlled plane flies because of the same basic design principles as a real aircraft there are factors which a real plane has to contend with that the model does not. At 30 000 feet the temperature is around –60 °C, wind speeds can be over 100 miles an hour, etc.

Models simplify a real situation while retaining essential features.

The creation of a mathematical model to describe a real situation can be summarised very broadly by the diagram below:

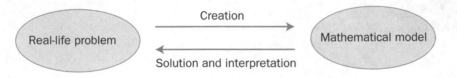

It is usually necessary to repeat this process a number of times, improving or extending the model to make it more realistic.

The accuracy of the model is assessed through a process of **validation**, where the answers produced by the model are compared with what is observed in reality. This will often involve the use of statistics. The results will then be used to revise the model.

A more realistic diagram of the modelling process is shown below:

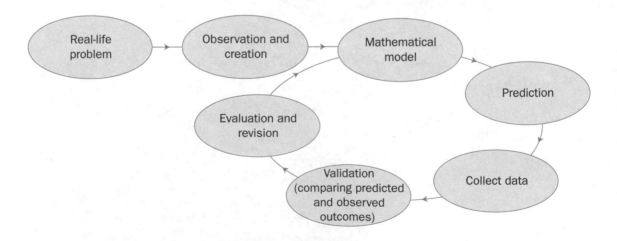

Some people prefer to think of the modelling process in a list:

Stage 1 A real-life problem is recognised.
Stage 2 A statistical model is devised.
Stage 3 The model is used to make predictions.
Stage 4 Experimental or observational data are collected.
Stage 5 Comparisons are made against the model.
Stage 6 Statistical techniques are used to test how well the model
         describes the real-life problem.
Stage 7 The model is refined.

Exercise 2.1

1 You are to design a model plane.
   Write down the factors which you think you would need
   to consider when deciding how powerful the engine will
   need to be.

2 You are to design a new washing machine.
   Write down the factors which you think the user should be
   able to control in the programme cycle.

3 You have to cross a busy road where there are no traffic lights
   or other crossing places.
   What factors should you take into account in deciding whether
   to cross the road at a particular time?

4 A delivery company wants to model the costs of fulfilling
   an order. The supervisor creates a list of the important factors
   which need to be in the model:

   ○ locating and packing the items in the order
   ○ vehicle and driver costs of making the delivery
   ○ administration and billing

For each of these, identify at least two variable quantities and
explain how you would measure or estimate them.

S1

Some situations contain elements of uncertainty, and this can pose problems when trying to model their behaviour.

Consider the following examples.

1 An airline has 143 passengers booked on the last flight of the day from London to Edinburgh.

- If the airline has to cancel the flight, how many of those passengers will want to travel to Edinburgh the next day?
- For how many people will the airline have to provide overnight accommodation?

It is not possible to know in advance exactly how many of the 143 passengers will want to travel the next day, and how many will want overnight accommodation. However, if the airline has some historical data on what has happened on previous occasions in similar circumstances, it can start making contingency plans.

Probability distributions will help you understand what happens in these situations.

2 A rare, but very serious, disease occurs in only one in 10 000 people. There is a screening test which gives a positive result in 99% of cases where the subject has the disease and gives a negative result in 99% of cases where the subject does not have the disease, i.e. it is 99% correct in both situations.

- How reasonable is it to tell a patient whose test comes back positive that they have this serious disease?

For every one person whose positive result occurred because they had the disease, there are likely to be about 100 people who gave a positive result without having the disease, because there are so many more people who do not have the disease. So it would not be reasonable to tell someone that the positive test alone means they have the disease.

You will look again at the case of the screening test in detail on page 69.

Conditional probability will help you understand the true implications of a positive result in a situation like this.

3   There is substantial evidence that shows that pupils in large classes have better examination results than pupils in small classes.

○   Would it be reasonable for the government to decide that all pupils should be taught in large classes in an effort to improve educational standards?

It would probably not be a good idea – schools put pupils who are more capable academically into larger classes, and their academic ability is the reason they have better results.

This is an example of where simply observing what is going on instead of conducting a designed statistical experiment could be very misleading.

4   James and Jean are twins who pass their driving test on the same day. Their father gives them each an identical car.

○   Would the cost of their insurance be the same?

Insurance companies ask customers for information about a range of factors which they think make a difference in the likelihood of people making a claim, or in the size of claim which might be made. Young males will be charged substantially more for insurance than females of the same age if all other factors are the same.

So James may find that his premium is 25% higher than Jean's.

HH LEARNING CENTRE
HARROW COLLEGE

Exercise 2.2

1 You are setting up an online book ordering service. One of the partners wants the website to make suggestions for other books a customer might like to read, based on what they have ordered and what other customers have done. What information could you use to achieve this?

You may like to go to websites offering online ordering with this feature and try to see how they do it.

2 Imagine you are a member of a jury which has just found a surgeon guilty of negligence in performing a knee operation on a young man. The damage to the knee means that the patient will not be able to play soccer again. You have to decide the level of compensation to award to the young man. What factors should you take into account in making that decision?
Three possible cases are given which may give you some ideas.

- Ali is 17. He left school last year and works in an office as a computer operator, and enjoys playing soccer with his friends at the leisure centre.
- Wayne is 20. Before the operation he was playing Premiership soccer and was an established international.
- Milan is 16. Before the operation he hoped to have a trial with a Premiership club; scouts from three clubs had watched him, but no one had invited him for a trial yet.

3 A mobile phone company is proposing to build a base station near your home, and you wonder whether there are any health risks. Write down any information you feel you would like to know to help you, and any difficulties you think there might be in reaching a conclusion to this question.

4 An outline of the stages in developing a statistical model are listed below with stages 1 and 6 missing.

Stage 1
Stage 2   A statistical model is devised.
Stage 3   The model is used to make predictions.
Stage 4   Experimental or observational data is collected.
Stage 5   Comparisons are made against the model.
Stage 6
Stage 7   The model is refined.

Suggest what stages 1 and 6 might be.

1 Explain briefly why statistical models are used when
   attempting to solve real-world problems.   [(c)Edexcel Limited 2002]

2 Statistical models can be used to describe real-world problems.
   Explain the process involved in the formulation of a statistical
   model.

3 Describe the role of simulation in testing statistical models.

4 Explain what you understand by a statistical model.

5 Describe two aspects of statistics which are commonly used in
   building a statistical model.   [(c)Edexcel Limited 2002]

2

Exit →

Summary

Refer to

- Mathematical models are used to represent many situations, for example:
  - how the space shuttle will behave under different conditions
  - how much cooling liquid a high speed drill needs to have in its system.

  2.1

- Modelling is a process which normally involves going through several cycles of evaluation and improvement until a satisfactory model is found. However, a model will always only be an approximation to the real situation.

  2.1

- Statistical models are models of situations where uncertainty is built in, for example:
  - Each year toy manufacturers hope to produce that year's 'must have' toy. However, manufacturing and distributing the toys requires a considerable lead time, so how do they decide how many toys to make?
  - In the event of a 'flu pandemic breaking out, what measures should the government take in order to limit the spread of the disease, and how much money should they spend in stockpiling vaccines in case they are needed?

  2.2

- Evaluation of statistical models is difficult because real-life outcomes will normally be different each time. Repeated simulations are needed to try to understand the likely range of outcomes in reality.

  2.2

Links

Loyalty cards: the new area of data mining made significant fortunes for the mathematicians and statisticians who developed the techniques allowing companies to target advertising effectively at individuals based on the information gathered through the use of loyalty cards.

In the financial services industry, derivatives trading is now very big business, and much of this relies very heavily on sophisticated statistical modelling.

S1

Representing and analysing data

This chapter will show you how to
- represent data in stem and leaf diagrams, histograms and boxplots
- calculate measures of centre and spread
- describe distributions
- make comparisons between distributions.

Before you start

You should know how to:

1 Construct basic statistical graphs.

2 Substitute values into algebraic expressions.

3 Find the mean and the upper and lower quartiles of a frequency distribution.

Check in

1 The heights of 30 children in Year 3 are given in the table.

Height (cm)	110–120	120–125	125–130	130–135
Frequency	5	8	10	7

Draw a histogram for the data.

2 **a** If $y = \frac{x - 22.5}{10}$, find y when $x = 42.5$

b If $y = 8x + 7.3$ find y when $x = 5$

3 Find the quartiles of this distribution of broad bean pods.

No. of beans per pod	1	2	3	4	5
Frequency	2	5	8	6	4

S1

Comparing two sets of data can be done in an effective way visually by using a back-to-back stem and leaf diagram.

The heights of samples of two types of plants are represented in this stem and leaf diagram.

	Type A								Type B						
						2	3	5	7	7					(4)
(2)					7	4	3	2	2	4	8	8	9		(6)
(6)	9	7	7	6	3	1	4	1	4	5	5	7			(5)
(5)		8	5	4	4	2	5	2	3						(2)
(3)				5	3	3	6								

Key 1 | 4 | 5 means 41 cm for type A and 45 cm for type B

Plants of type A are taller on average than plants of type B, and the two types of plant have similar variability in their heights.

When comparing distributions, you should comment on average and spread. The comments need to be in the context of the data.

Exercise 3.1

1 The data below record the masses, in grams, of two samples of 35 plums of different types.

Type A

2	5	(1)
3	0 2 4	(3)
3	5 5 7 7 9	(5)
4	1 2 2 2 2 4 4 4 4	(9)
4	5 5 6 6 6 7 7 7 7 8	(10)
5	0 1 2 2	(4)
5	7 8 9	(3)

Key: 2 | 5 means 25 g

Type B

4	4	(1)
4	5 6 7 7 8 8 8 9 9 9 9	(11)
5	0 0 0 0 1 1 1 1 1 2 2 3 3 4	(14)
5	5 5 6 6 7 7 8 9 9	(9)

Key: 4 | 4 means 44 g

a For each type, find the median and interquartile range of the masses.

b Draw a back-to-back stem and leaf diagram to show these data in a single diagram.

c Compare the characteristics of the two types.

d If you were a fruit-grower, which of the two types would you plant? Give a reason.

S1

2 A group of Year 12 students measured their pulse rates at the
   start of a statistics lesson. They then did five minutes of moderate
   exercise and measured their pulse rates again:

	Before exercise		After exercise	
(3)	9 8 6	6		
(4)	7 5 4 2	7		
(4)	8 6 5 1	8	4 5 7 7 8	(5)
(3)	5 3 2	9	1 3 4 4 6 9	(6)
(1)	0	10	0 2 5 5	(4)

Key 2 | 9 | 1 means 92 beats per minute before exercise and 91 beats per minute after exercise

Compare the pulse rates of the group before and after exercise.

3 Samples of the weights of dunnocks are taken in January and
   April.

A dunnock is a small British songbird.

	January		April	
(1)	5	18	6	(1)
(4)	9 6 2 1	19	1 4 5 7 8	(5)
(7)	9 8 8 5 5 2 0	20	0 2 4 4 5 8 9	(7)
(8)	7 6 5 3 3 3 2 2	21	0 1 2 5 5 5 6 7 8 9	(10)
(8)	9 9 9 4 4 2 1 0	22	0 0 2 3 3 4 8 9	(8)
(5)	4 3 2 1 0	23	4	(1)
(1)	2	24	5 7	(2)
(2)	2 0	25		

Key: 6 | 19 | 7 means 19.6 g for January and 19.7 g for April.

Compare the weights of dunnocks in January and in April.

4 The wingspans of a sample of male and female sparrowhawks
   were measured, and are summarised in the diagram below.

	Male		Female	
(2)	9 8	18		
(11)	4 4 4 3 3 3 2 2 2 1 1	19		
(26)	9 8 8 8 7 7 7 7 7 6 6 6 6 6 6 5 5 5 5 5 5 5 5 5 5 5	19		
(9)	4 3 2 2 2 0 0 0 0	20		
		20		
		21		
		21		
		22		
		22	5 5 6 8	(4)
		23	0 2 2 3	(4)
		23	5 5 6 9	(4)
		24	0	(1)
		24		(0)
		25	1	(1)

Key: 4 | 20 means 20.4 cm and 22 | 5 means 22.5 cm.

Compare the wingspans of male and female sparrowhawks.

Boxplots typically show only five points from the distribution –
the top and bottom of the range, the median and two quartiles.

This pair of boxplots show the heights of two groups of plants.

Boxplots can run vertically, as
shown here, or horizontally, as
shown at the foot of the page.

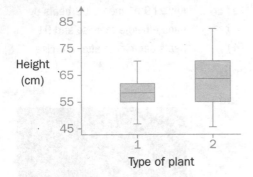

Group 1 plants are much more consistent in height than those in
group 2, but the plants in group 2 tend to be taller.

Outliers

Outliers are values that are uncommonly large or small compared
to the rest of the data.

The usual definition of an outlier is a value which is more than
$1.5 \times$ IQR above the UQ (upper quartile) or below the LQ
(lower quartile). These limits are known as **fences** and values
outside them can be identified individually.

$IQR = Q_3 - Q_1$

$x - Q_3 > 1.5 \times (Q_3 - Q_1)$

$Q_1 - x > 1.5 \times (Q_3 - Q_1)$

This example shows outliers in boxplots. There is a single value
at 3 and the next smallest value is 9. The largest value is 19.

The fences are at $11 - 1.5 \times 4 = 5$ and at $15 + 1.5 \times 4 = 21$,
but there are no values larger than 19, so the whisker
stops at the largest value.

$IQR = Q_3 - Q_1$
$\quad = 15 - 11$
$\quad = 4$

The smallest value above the lower fence is 9, so the whisker
stops at 9, with the outlier at 3 still showing.

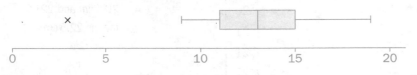

Exercise 3.2

1 The two boxplots show salaries in a college in the USA.

a The first shows the salaries paid during the 1991–1992 academic year, separately for males and females. Is there any difference in the treatment of men and women in this college?

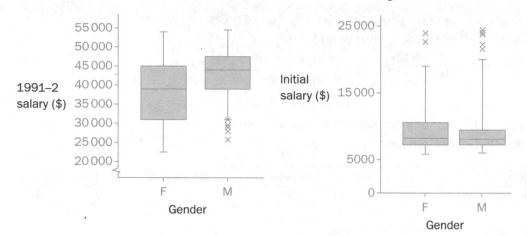

b The second shows the staff members' initial salary, in whatever year they joined the staff, again shown separately for males and females. Does the information contained in this diagram alter any views you formed from the first dataset? (Explain how, if it does.)

2 The box and whisker plots on the right show the average daily temperatures in two holiday resorts in July. Compare the two resorts as holiday destinations, and state any other information you might wish to know.

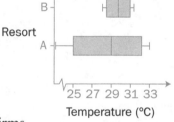

3 The quartiles of the number of employees in random samples of firms in 1990 and in 1995 are given in a table below, together with any outliers which are more than 1.5 × the interquartile range beyond the quartile.

	Min	LQ	Median	UQ	Max	Outliers
1990	1	6	18	42	145	106, 131, 145
1995	1	3	13	35	160	95, 160

a Construct a boxplot for each year.

b Compare and comment on the two distributions.

In a histogram, the area of each bar is proportional to the frequency in the interval.

> When the intervals are not of equal width the height of the bar is the frequency density and it must be scaled:
>
> $$\text{Frequency density} = \frac{\text{frequency}}{\text{interval width}}$$

EXAMPLE 1

The heights of the children in a school are measured correct to the nearest centimetre and are summarised in the table.

Height (cm)	120–129	130–139	140–144	145–149	150–154	155–159	160–169	170–179
Frequency	60	80	50	93	77	67	72	54

Draw a histogram to represent this data.

Height (cm)	120–129	130–139	140–144	145–149	150–154	155–159	160–169	170–179
Frequency	60	80	50	93	77	67	72	54
Class width	10	10	5	5	5	5	10	10
Frequency density	6	8	10	18.6	15.4	13.4	7.2	5.4

Draw the histogram:

The interval endpoints are 119.5, 129.5, 139.5, 144.5, 149.5, 154.5, 159.5, 169.5 and 179.5

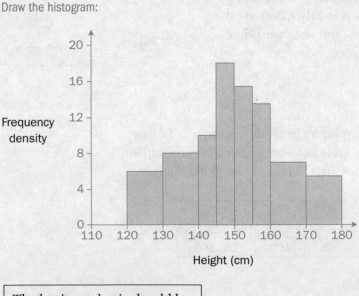

The horizontal axis should be drawn as a linear scale showing the variable.

Relative frequency histograms

In a **relative frequency histogram**, the total area is defined to be one square unit.

Continuous probability distributions are defined like this – see page 167.

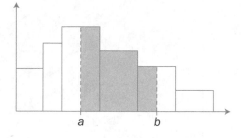

The proportion of data which lies between any two values will be the area between those two values.

To calculate the heights of the bars for a relative frequency histogram, you divide the frequency densities by the total frequency – this will automatically scale the total area to be 1.

Example 1 then looks like this.

Total frequency = 60 + 80 + 50 + 93 + 77 + 67 + 72 + 54 = 553

Height	120–129	130–139	140–144	145–149	150–154	155–159	160–169	170–179
Frequency density	6	8	10	18.6	15.4	13.4	7.2	5.4
F.d ÷ total frequency (3 d.p.)	0.011	0.014	0.018	0.034	0.028	0.024	0.013	0.010

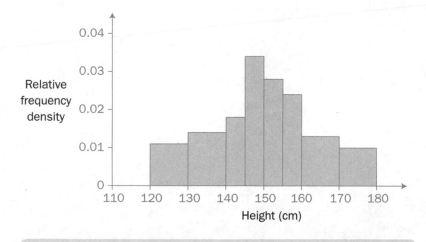

Total frequency
= 553
Height of first bar
= 6 ÷ 553 = 0.0108...
and so on.

$$\text{Relative frequency density} = \frac{\text{frequency density}}{\text{total frequency}}$$

S1

EXAMPLE 2

SI

The ages of members of an amateur dramatic society are summarised in the table below.

Age (years)	25–34	35–44	45–49	50–54	55–59	60–74
Frequency	6	9	13	9	7	6

a Draw a relative frequency histogram to display this information.

b Calculate an estimate of the number of members aged between 40 and 52, inclusive.

a Since these are ages, the ends of the first interval are at 25, 35, etc. A table is the best way to show the calculations for either sort of histogram.

Ages	Frequency	Interval width	Frequency density	Relative frequency density
25–34	6	10	0.6	0.012
35–44	9	10	0.9	0.018
45–49	13	5	2.6	0.052
50–54	9	5	1.8	0.036
55–59	7	5	1.4	0.028
60–74	6	15	0.4	0.008
Total	50			

Use the tabulated values to draw a histogram:

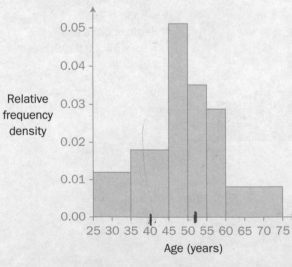

b To estimate the number of members aged between 40 and 52, include half of the 35–44 interval, all of 45–49 and $\frac{3}{5}$ of the 50–54 interval.

∴ Estimated number of members aged between 40 and 52 =
$0.5 \times 9 + 13 + 0.6 \times 9 = 22.9$ or 23

Exercise 3.3

1 The table gives the length in cm of a number of different varieties of plant.

Length (cm)	20–50	–60	–65	–70	–80	–90	–110
Number of plants	27	18	16	15	22	14	14

Draw a histogram for these data.

2 The table gives the daily protein intake of a sample of 130 patients in a hospital.

Protein (grams)	0–15	–25	–30	–35	–40	–45	–50	–55	–65
Number of people	18	14	11	16	21	18	15	11	6

Draw a histogram for these data.

3 A question on a survey asked children how long their journeys to school had taken on a particular day.
A summary of the results is shown in the table.

Time (min, correct to nearest min)	1–15	16–25	26–35	36–55	56–80
Number of children	115	46	36	22	14

Illustrate these data using an appropriate diagram.

4 The table shows information about the salaries paid to employees in a company.

Draw a histogram to represent these data.

Salary	Frequency
£0 $< x \leqslant$ £10 000	7
£10 000 $< x \leqslant$ £15 000	82
£15 000 $< x \leqslant$ £20 000	45
£20 000 $< x \leqslant$ £25 000	24
£25 000 $< x \leqslant$ £30 000	13
£30 000 $< x \leqslant$ £50 000	4

5 The length of time students take to complete a mathematical puzzle is summarised in the table.

Time, t (min)	0–2	2–5	5–8	8–12	12–20
Number of pupils	6	8	15	14	7

a Draw a relative frequency histogram to represent these data.

b Calculate an estimate of the number of pupils who took more than 10 minutes, but not more than 15 minutes, to complete the puzzle.

3.4 Quantiles

You can divide data into parts using quantiles.

You have already met quartiles in Chapter 1.
Other common quantiles are

deciles   these divide the dataset into 10 parts
percentiles   these divide the dataset into 100 parts

Quantiles are calculated in the same way as quartiles.

For a dataset with n values $[x_1, x_2, \ldots, x_n]$, to find the 7th decile, calculate $\frac{7}{10}n$.

For example, $n = 23$
$$\frac{7}{10} \times 23 = 16.1$$
The 7th decile is x_{17}.

If $\frac{7}{10}n$ is an integer, r, then the 7th decile is the midpoint of x_r and x_{r+1}.

If $\frac{7}{10}n$ lies between r and $r + 1$ then the 7th decile is x_{r+1}.

EXAMPLE 1

Here is a set of data.

5	6	7	7	8	8	8	9
10	11	14	15	15	16	17	19
21	22	24	27	28	29	31	31
32	34	36	37	38	39	41	42
43	46	47	49	51	53	54	57

Find   **a** the 6th decile
          b the 37th percentile.

---

a $n = 40$
$$\frac{6}{10} \times 40 = 24$$

so the 6th decile is the midpoint of the 24th and 25th values

6th decile $= \frac{(31 + 32)}{2} = 31.5$

b $\frac{37}{100} \times 40 = 14.8$

so the 37th percentile is the 15th value in the list.

The 37th percentile $= 17$

The 6th decile can be written as D_6.
The 37th percentile can be written as P_{37}.

Grouped frequency distributions

When data is given in a grouped frequency table, you can use interpolation to estimate the value of quantiles.

EXAMPLE 2

A question on a survey asks children how long their journey to school took on a particular day. A summary of the results is shown below.

Time (minutes, correct to the nearest minute)	Number of children
1–15	115
16–25	46
26–35	36
36–55	22
56–80	14

Calculate estimates of   **a**   the upper quartile

b   the 95th percentile for these data.

---

a $n = 233$, and Q_3 is the $\left(233 \times \frac{3}{4}\right)$th value $= 174.75$th value

Rewrite the table with cumulative frequencies:

115, 161, 197, 219, 233

174.75 is between 161 and 197, so the upper quartile lies in the 26–35 interval.

174.75 − 161 = 13.75 and there are 36 children in the interval, so the estimate will be $\frac{13.75}{36}$ of the way through the interval.

The interval starts at 25.5 and is of width 10 minutes so the estimated upper quartile is

$25.5 + \frac{13.75}{36} \times 10 = 29.3194\ldots = 29.3$ minutes

```
  161      174.75          197
   |---------|--------------|
  25.5       ?             35.5
```

b P_{95} is the $\left(233 \times \frac{95}{100}\right)$th value $= 221.35$th value

221.35 is above 219 so the 95th percentile lies in the last interval.

221.35 − 219 = 2.35 and there are 14 children in the last interval which has width 25 minutes.

$55.5 + \left(\frac{2.35}{14}\right) \times 25 = 59.696\ldots = 59.7$ minutes

```
 219  221.35              233
  |----|-------------------|
 55.5  ?                  80.5
```

You may prefer to use a formula for interpolation:

Estimated value of quantile

$$= \text{lower class boundary} + \left(\frac{Q - \text{cum. freq. at start of interval}}{\text{interval frequency}} \right) \times \text{interval width}$$

where Q is the position of the quantile required.

Exercise 3.4

1 A group of Year 12 students recorded their pulse rates when they were in a relaxed state.

53	54	55	55	57	58	59	59	59	62	62	62
62	62	63	63	63	64	64	65	65	67	67	67
68	68	69	69	70	71	71	73	75	79	80	83

Calculate:

a the lower quartile

b the 7th decile

c the 84th percentile of these pulse rates.

2 A group of 45 women go on a skiing holiday. Their ages are listed below.

17	17	17	19	19	19	20	22	22	22	23	23	24	25	25
25	26	26	26	26	27	27	28	28	28	29	29	29	30	32
32	32	34	34	35	35	35	35	36	37	38	38	44	45	48

Calculate:

a the median

b the 3rd decile

c the 19th percentile of these pulse rates.

3 The table shows information about the salaries paid to employees in a company.

Calculate estimates of:

a the lower quartile

b the 8th decile.

Salary	Frequency
£0 < x ⩽ £10 000	7
£10 000 < x ⩽ £15 000	82
£15 000 < x ⩽ £20 000	45
£20 000 < x ⩽ £25 000	24
£25 000 < x ⩽ £30 000	13
£30 000 < x ⩽ £50 000	4

4   The length of time pupils take to complete
    a mathematical puzzle is summarised in the table.

Time, t (minutes)	Number of pupils
$0 < t \leqslant 2$	6
$2 < t \leqslant 5$	8
$5 < t \leqslant 8$	15
$8 < t \leqslant 12$	14
$12 < t \leqslant 20$	7

Calculate estimates of:

a   the upper quartile

b   the 61st percentile.

5   At a metro station, a regular passenger times (in seconds) how
    long he has to wait for a train to arrive once he reaches the
    platform. These data are listed below.

    87   42   0   62   124   0   58   37   74   94
    182   23   17   62   29   17   82   54   0   45

Find:

a   the lower quartile

b   the 4th decile

c   the 34th percentile.

6   The length of time, in minutes, that customers spend in a
    coffee shop is recorded. The results are shown below.

    17, 15, 9, 31, 33, 41, 8, 14, 13, 22, 27, 43, 32, 14

Find:

a   the 3rd quintile

b   the 27th percentile

c   the difference between the 1st and 9th deciles.

Quintiles divide the data into
five parts.

3.5 Variance and standard deviation

The mean of a set of data is given by the formula:

$$\text{Mean} = \bar{x} = \frac{\sum x}{n} \quad \text{or} \quad \frac{\sum xf}{\sum f}$$

Remember Σ means 'the sum of'.

where x is each data value, f is the frequency, n is the number of data values.

You have met two measures of spread, the range and the interquartile range. Each of these measures only uses two values. Another measure of spread, which uses all the data values, is the variance.

The variance of a set of data is defined as the average of the squared distances from the mean.

$$\text{Variance} = \sum \frac{(x - \bar{x})^2}{n} \quad \text{or} \quad \frac{\sum (x - \bar{x})^2 f}{\sum f}$$

$$= \frac{\sum x^2}{n} - \bar{x}^2 \quad \text{or} \quad \frac{\sum x^2 f}{\sum f} - \bar{x}^2$$

The version of the formula without brackets is generally easier to work with.

$$\sum \frac{(x - \bar{x})^2}{n}$$

$$= \sum \left(\frac{x^2 - 2x\bar{x} + \bar{x}^2}{n} \right)$$

$$= \sum \left(\frac{x^2}{n} \right) - 2\bar{x} \sum \left(\frac{x}{n} \right) + \bar{x}^2 = \sum \left(\frac{x^2}{n} \right) - 2\bar{x}^2 + \bar{x}^2$$

$$= \frac{\sum x^2}{n} - \bar{x}^2$$

You will not need to reproduce this derivation.

$\sum \bar{x}^2 = n\bar{x}^2$ and $\sum \frac{x}{n} = \bar{x}$

The standard deviation is the square root of the variance. It is a measure of the spread of the data, and has the same units as the data.

Find the mean and standard deviation of 11, 13, 14, 16 and 18 cm.

$\sum x = 11 + 13 + 14 + 16 + 18 = 72$ cm

so the mean $= \frac{72}{5} = 14.4$ cm

$\sum x^2 = 11^2 + 13^2 + 14^2 + 16^2 + 18^2 = 1066$ cm^2

so variance $= \frac{1066}{5} - 14.4^2 = 5.84$ cm^2

and the standard deviation $= \sqrt{5.84} = 2.42$ cm (3 s.f.)

Using the other form:
$\sum (x - \bar{x})^2$

$= 3.4^2 + 1.4^2 + 0.4^2 + 1.6^2 + 3.6^2$

$= 29.2$

so the variance $= \frac{29.2}{5} = 5.84$

You can use your calculator's statistical functions to obtain the mean and standard deviation. Your calculator will normally have two versions, which give slightly different answers.

The population standard deviation is the one you want, usually shown as σ or σ_n on a calculator. The other is sometimes referred to as a sample standard deviation, usually shown as s or σ_{n-1}.

EXAMPLE 2

The number of children, x, in each of the families of a class of 30 pupils was recorded. Find the mean and standard deviation of the number of children in a family for that class given that $\sum x = 66$ $\sum x^2 = 165$.

Mean $= \dfrac{66}{30} = 2.2$ children

variance $= \dfrac{165}{30} - 2.2^2 = 0.66$

Standard deviation $= \sqrt{0.66} = 0.81$ children

EXAMPLE 3

The heights of 142 plants are measured and recorded in the table below where the class 65–70 means at least 65 cm and less than 70 cm tall.

Class	55–60	60–65	65–70	70–75	75–80	80–85	85–90
Frequency	2	11	37	54	28	9	1

Estimate the mean and variance of the heights of the plants.

You need to calculate the midpoints of each class, and a table of values is the simplest way to do the work required.

For a grouped frequency table you can only make an **estimate** of the mean.

Class	m	f	fm	fm^2
55–60	57.5	2	115	6612.5
60–65	62.5	11	687.5	42968.75
65–70	67.5	37	2497.5	168581.3
70–75	72.5	54	3915	283837.5
75–80	77.5	28	2170	168175
80–85	82.5	9	742.5	61256.25
85–90	87.5	1	87.5	7656.25
		$\sum f = 142$	$\sum fm = 10215$	$\sum fm^2 = 739087.5$

Then $\bar{x} = \dfrac{10\,215}{142} = 71.9$ and variance $= \dfrac{73\,9087.5}{142} - \left(\dfrac{10\,215}{142}\right)^2$

$$= 29.964\ldots$$

So the mean is 71.9 cm and the variance is 30.0 cm^2 (to 1 d.p.)

S1

EXAMPLE 4

The number of putts, x, a golfer takes on each of the 18 holes in a competition is recorded. Find the standard deviation of the number of putts he takes per hole, given that
$$\sum(x - \bar{x})^2 = 4.944$$

There were 18 holes, so

variance $= \dfrac{4.944}{18} = 0.2746\ldots$

Standard deviation $\sqrt{0.2746\ldots} = 0.524$ putts

Variance $= \sum \dfrac{(x - \bar{x})^2}{n}$

What does standard deviation actually mean?

A good visual idea of standard deviation is to imagine a histogram of the data. For a roughly symmetrical distribution, a spread of four standard deviations covers about the central 95% of the distribution.

You will learn about an important symmetrical distribution in Chapter 8.

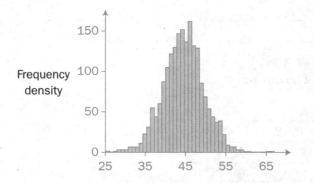

Here the bulk of the data is between 35 and 55 so four standard deviations is around 20, the standard deviation is about 5.

Exercise 3.5

1 For the following sets of data, calculate the mean and standard deviation. You should do this using the formulae, and check using your calculator.

a

12	17	11	8	6	18	14	17
11	15	16	18	9	15	20	14

b

x	4	5	6	7	8	9
f	12	18	35	28	16	9

2 The wingspans (in cm) of 352 great tits were measured in 2007, and the information is summarised in the table.

Wingspan (cm)	70	71	72	73	74	75	76	77	78	79
Frequency	4	19	53	77	80	75	53	30	12	6

Calculate the mean and standard deviation of the wingspan of the great tits.

3 The total head lengths for a sample of 69 treecreeper birds
are summarised in the table below.

Head length (mm)	295	297	300	305	310	312	315	320	325	328	330	335	340	345
Frequency	4	1	6	5	11	1	8	10	5	1	12	3	1	1

Calculate the mean and standard deviation of the head lengths
of the treecreepers.

4 The goldcrest is the UK's smallest songbird. The weights
(in g to the nearest 0.1 g) of 210 goldcrests were measured during
2006 and 2007. The information is summarised in the
table below.

Weight (g)	4.0–4.4	4.5–4.9	5.0–5.4	5.5–5.9	6.0–6.4
Frequency	3	45	149	11	2

Calculate estimates of the mean and standard deviation of the
weight of a goldcrest.

5 $\sum x = 75$, $\sum x^2 = 293$, $n = 30$. Find the mean and variance of x.

6 $\sum x = 2.1$, $\sum x^2 = 83.1$, $n = 7$. Find the mean and standard
deviation of x.

7 $\sum(x - \bar{x})^2 = 44.3$ $n = 16$ Find the variance of x.

8 $\sum(x - \bar{x})^2 = 19\,735.4$ $n = 37$ Find the standard deviation of x.

9 Give estimates of the mean and standard deviation of this
distribution.

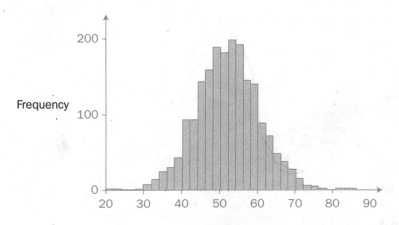

The diagram shows two sets of data.
Dataset 1 represents the ages in years of a group of people.
Dataset 2 represents the ages of the same people 15 years later.

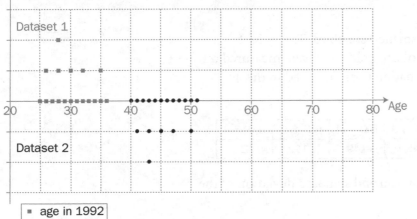

All the data values have shifted by 15, so
Mean of dataset 2 = (mean of dataset 1) + 15
There is no change to any measure of spread.

The next diagram shows a set of temperatures in Celsius (Dataset 1),
and the same temperatures in Fahrenheit (Dataset 2).

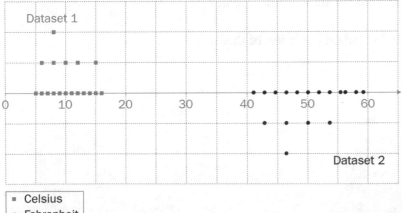

All of the Celsius data values have been multiplied by 1.8, then 32 has
been added.
Mean of dataset 2 = 1.8 × (mean of dataset 1) + 32
The degree of spread is unaffected by the +32.
Standard deviation of dataset 2 = 1.8 × (standard deviation of dataset 1)
So variance of dataset 2 = 1.8^2 × (variance of dataset 1)

If a set of data values X is related to a set of values Y so that $Y = aX + b$, then
- mean of $Y = a \times$ mean of $X + b$
- standard deviation of $Y = a \times$ standard deviation of X.
- variance of $Y = a^2 \times$ variance of X.

You can use this idea to transform or code a set of numbers. You might do this to make the numbers easier to work with.

Another use of coding is in standardising different data sets for comparison.

EXAMPLE 1

Here is a set of grouped data.

Class	m	f
55–60	57.5	2
60–65	62.5	11
65–70	67.5	37
70–75	72.5	54
75–80	77.5	28
80–85	82.5	9
85–90	87.5	1
		$\Sigma f = 142$

Use the coding $M = \frac{m - 57.5}{5}$ to find the mean and the variance of this data.

Extend the table:

Class	m	M	f	fM	fM^2
55–60	57.5	0	2	0	0
60–65	62.5	1	11	11	11
65–70	67.5	2	37	74	148
70–75	72.5	3	54	162	486
75–80	77.5	4	28	112	448
80–85	82.5	5	9	45	225
85–90	87.5	6	1	6	36
			$\Sigma f = 142$	$\Sigma fM = 410$	$\Sigma fM^2 = 1354$

The coded numbers M are easier than the uncoded numbers m.

Then $\bar{M} = \frac{410}{142} = 2.887\ldots$

and $\sigma_M^2 = \frac{1354}{142} - \bar{M}^2 = 1.19857\ldots = 1.20$

To find the mean and variance of the uncoded data
$\bar{m} = \bar{M} \times 5 + 57.5 = 2.887 \times 5 + 57.5 = 71.9$ (3 s.f.)
and $\sigma_m^2 = 5^2 \times \sigma_M^2 = 25 \times 1.19857\ldots = 30.0$ (3 s.f.)

Without coding, $\Sigma fm = 10\,215$
$\Sigma fm^2 = 739\,087.5$
$\bar{m} = \frac{10\,215}{142} = 71.9$

$\sigma_m^2 = \frac{739\,087.5}{142} - \left(\frac{10\,215}{142}\right)^2 = 30.0$

Coding gives the same answer but uses smaller numbers and fewer key presses.

Exercise 3.6

1 A set of observations $\{x\}$ is coded using $X = \frac{x - 62.5}{5}$.
$\bar{X} = 3.6$ Variance of $X = 5.3$
Calculate the mean and variance of the original set
of observations.

2 A set of observations $\{x\}$ is coded using $X = \frac{x - 1055}{10}$.
$\bar{X} = 8.2$ Variance of $X = 3.22$.
Calculate the mean and variance of the original set
of observations.

3 The temperatures (in °F) at a resort are measured at the same
time on eight successive Mondays:

50.0   53.6   52.7   55.4   55.4   57.2   59.9   62.6

a Calculate the mean of these temperatures (in °F).

b Convert each temperature into °C by the formula $C = \frac{F - 32}{\left(\frac{9}{5}\right)}$

c Calculate the mean of these temperatures (in °C).

d Check that the mean temperature in °F converts to the same
mean in °C.

4 Some information about the times taken by a bus to travel
from Nikki's home to her grandmother's house is shown.

Time	Frequency
$8 \leqslant x < 10$	5
$10 \leqslant x < 12$	14
$12 \leqslant x < 14$	3
$14 \leqslant x < 16$	2

a Calculate estimates of the mean and variance of the times
taken by the bus.

b Code this data by using $X = \frac{x - 9}{2}$.

c Find estimates of the mean and variance of the coded data.

d Check that the mean of $x = 2 \times$ mean of $X + 9$
and that the variance of $x = 2^2 \times$ variance of X.

5 The table shows information about the salaries paid to
   employees in a company.

Salary	Frequency
£0 < x ⩽ £10 000	12
£10 000 < x ⩽ £15 000	63
£15 000 < x ⩽ £20 000	32
£20 000 < x ⩽ £25 000	21
£25 000 < x ⩽ £30 000	8
£30 000 < x ⩽ £50 000	4

a Code the data by using $X = \dfrac{x - 5000}{2500}$.

b Calculate estimates of the mean and variance of the
coded data.

c Use your answers to part **b** to give estimates of the mean
and variance of the salaries in the company.

6 The table shows information about the daily times spent with
   patients by a nurse in a care home during one month.

Time, t (minutes)	Frequency
10 < t ⩽ 15	312
15 < t ⩽ 20	479
20 < t ⩽ 25	243
25 < t ⩽ 30	119

a Code the data by using $T = \dfrac{t - 12.5}{5}$.

b Calculate estimates of the mean and variance of the
coded data.

c Use your answers to part **b** to give estimates of the mean
and variance of the time spent with a patient.

S1

Skewness is the term used to describe the lack of symmetry in a distribution.

Histograms and boxplots can show skewness quite well, particularly where data is not symmetric about the median.

A **positively skewed** distribution has:
- a long tail to the right
- $Q_3 - Q_2 > Q_2 - Q_1$
- mode < median < mean.

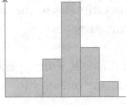

A **negatively skewed** distribution has:
- a long tail to the left
- $Q_3 - Q_2 < Q_2 - Q_1$
- mean < median < mode.

A **symmetric** distribution has:
- equal length tails
- $Q_3 - Q_2 \approx Q_2 - Q_1$
- median $\approx$ mean.

There are some common measures of skewness that quantify skewness as well as giving the sign (positive or negative).

Here are two such measures:

$$\frac{3 \times (\text{mean} - \text{median})}{\text{standard deviation}} \quad \text{and} \quad \frac{\text{mean} - \text{mode}}{\text{standard deviation}}$$

You do not need to learn these measures.

Also, the **quartile skewness coefficient** is defined by

$$\frac{(Q_3 - Q_2) - (Q_2 - Q_1)}{Q_3 - Q_1} = \frac{Q_3 - 2Q_2 + Q_1}{Q_3 - Q_1}$$

All these measures allow comparisons to be made between different distributions.

EXAMPLE 1

Here are the summary statistics for some data:
Mean = 10.7, standard deviation = 9.3,
lower quartile = 3.6, median = 8.6,
upper quartile = 15.4 and mode = 2.7
The boxplot is shown below.

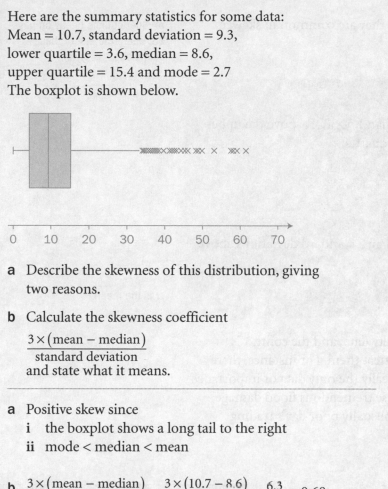

a Describe the skewness of this distribution, giving two reasons.

b Calculate the skewness coefficient

$$\frac{3 \times (\text{mean} - \text{median})}{\text{standard deviation}}$$

and state what it means.

a Positive skew since
   i the boxplot shows a long tail to the right
   ii mode < median < mean

b $\dfrac{3 \times (\text{mean} - \text{median})}{\text{standard deviation}} = \dfrac{3 \times (10.7 - 8.6)}{9.3} = \dfrac{6.3}{9.3} = 0.68$

The positive value indicates that the skew is positive.

Note that in the last example you could have worked out other skewness coefficients:

$$\frac{\text{mean} - \text{mode}}{\text{standard deviation}} = \frac{10.7 - 2.7}{9.3} = \frac{8}{9.3} = 0.86$$

$$\frac{(Q_3 - Q_2) - (Q_2 - Q_1)}{Q_3 - Q_1} = \frac{Q_3 - 2Q_2 + Q_1}{Q_3 - Q_1} = \frac{15.4 - 2 \times 8.6 + 3.6}{15.4 - 3.6}$$

$$= \frac{1.8}{11.8} = 0.15$$

The numerical value of each skewness coefficient is different.
To compare distributions it is therefore important that you use the same measure.

Outliers

Outliers are extreme values, and they are common in skewed distributions.

You met outliers briefly on page 28, where the first of these definitions was given.

An outlier is often defined as:

o any value x which is more than $1.5 \times$ IQR above the upper quartile or below the lower quartile

$$x - Q_3 > 1.5 \times (Q_3 - Q_1)$$
$$Q_1 - x > 1.5 \times (Q_3 - Q_1)$$

Another definition is:

o any value x which is more than 2 standard deviations above or below the mean, i.e.

$$\left| \frac{x - \mu}{\sigma} \right| > 2$$

μ is the mean.

Outliers may be the result of faulty data, and the context is important in deciding how you treat them. For instance, there are times when the outliers are really the only data of importance e.g. unusually high tides can cause tremendous flood damage, or in the financial markets an unusually poor day's trading can trigger a crisis of confidence.

Exercise 3.7

1 Summary statistics are given for three groups in the table.

   a Calculate the coefficient of skewness

   $\dfrac{3 \times (\text{mean} - \text{median})}{\text{standard deviation}}$ for each of the groups.

   b Which of the three distributions is most skewed by this measure?

	Group A	Group B	Group C
Mean	31.7	89.2	54.0
Standard deviation	6.4	11.3	7.2
Median	32.1	95.3	56.5
Lower quartile	21.5	81.6	47.2
Upper quartile	41.2	102.5	62.7
Mode	29.1	96.1	55.8

2 Summary statistics are given for three groups in the table.

   a Calculate the coefficient of skewness

   $\dfrac{Q_3 - 2Q_2 + Q_1}{Q_3 - Q_1}$ for each of the groups.

   b Which of the three distributions is most skewed by this measure?

   c Does the coefficient of skewness $\dfrac{\text{mean} - \text{mode}}{\text{standard deviation}}$

   give the same order of skewness for these groups?

	Group A	Group B	Group C
Mean	84.1	79.3	51.1
Standard deviation	6.2	17.2	8.3
Median	85.1	84.2	52.1
Lower quartile	79.3	71.1	41.0
Upper quartile	90.1	93.2	59.3
Mode	86.2	85.3	53.1

S1

3 The ages of applicants for mortgages are recorded by an estate agency. The results are shown below.

25, 29, 27, 32, 45, 34, 26, 28, 30, 42, 26, 51, 29, 27, 33, 27

For these data:

a calculate the mean

b draw a stem and leaf diagram

c find the median and the quartiles.

An outlier is an observation that falls either 1.5 × (interquartile range) above the upper quartile or 1.5 × (interquartile range) below the lower quartile.

d Determine whether or not any items of data are outliers.

e On graph paper draw a boxplot to represent these data. Show your scale clearly.

f Comment on the skewness of the distribution of ages of applicant for mortgages. Justify your answer.

4 At a metro station, a regular passenger times (in seconds) how long he has to wait for a train to arrive once he gets to the platform. These data are listed below.

87 42 0 62 124 0 58 37 74 94
182 23 17 62 29 17 82 54 0 45

a Find the median and interquartile range of the waiting times.

An outlier is an observation that falls either 1.5 × (interquartile range) above the upper quartile or 1.5 × (interquartile range) below the lower quartile.

b Draw a boxplot to represent these data, clearly indicating any outliers.

c Comment on the skewness of these data. Justify your answer.

d Explain how a zero waiting time occurs.

S1

3.8 Comparing distributions

There are various measures of average that you can use to compare distributions, and you are not expected to calculate all of them.

The mean uses all data values, but can be distorted by outliers.

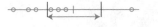

mean

The median is the middle value, and is less influenced by outliers.

median

The modal class has the highest frequency density. In a histogram, it is the interval with the tallest block.

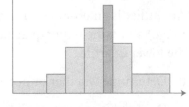

Similarly there are various measures of spread that can be used to compare distributions.

The standard deviation is calculated from the mean. It can therefore be distorted by outliers.

The interquartile range is used with the median. It concerns the middle 50%, and is unaffected by outliers.

The range is the difference between the highest and lowest values, so it is affected by extreme values.

If you are comparing distributions:
o make the comparison in context
o make reference to both the average and the spread.

Only use a single measure of average and a single measure of spread.

EXAMPLE 1

The times taken by pupils from three schools (A, B, C) to complete a mathematical challenge are summarised in the boxplots below. Compare the performance of the pupils in the three schools.

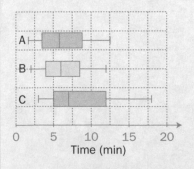

Pupils in schools A and B were generally faster than school C – they took less time and there was less variability in the times they took. School A had pupils with the shortest times but also had a few pupils with longer times than any of school B's.

You could say a little more, but be careful not to get into minute detail.

SI

Exercise 3.8

1 A group of pupils took their pulse rates at the end of a
   mathematics lesson. The mean was 64.2 and the standard
   deviation was 6.5.

   At the end of the next lesson, which was PE, they took their
   pulse rates again. The mean was now 71.6 and the standard
   deviation was 9.1.

   Compare the pulse rates before and after the PE lesson.

2 A new diet is claimed to increase the speed of weight loss.
   A random sample of 12 volunteers followed the diet and
   recorded their weight loss in one week.
   The results were (in kilograms):

   1.3, 1.1, 0.5, 1.3, 1.5, 1.4, 0.8, 1.2, 1.0, 0.8, 1.2, 1.1

   a Calculate the mean and variance of the weight loss with
   this diet.

   A traditional diet has a weight loss with a mean of 0.8 kg and
   variance 0.1 kg^2.

   b Compare the weight loss from the two diets.

3 The boxplots show the lengths (in cm) of a certain type of
   plant found in two gardens.

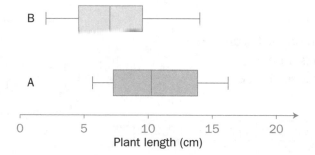

   Compare the plants in gardens A and B.

1 Air traffic control at an airport records the delays in arrival times of flights into the airport. On one day, 40% of all flights had no delay, the longest delay was 44 minutes, and half of all flights had delays of no more than 4 minutes. A quarter of all delays were at least 18 minutes, but only one was more than 25 minutes.

An outlier is an observation that falls either 1.5 × (interquartile range) above the upper quartile or 1.5 × (interquartile range) below the lower quartile.

a On graph paper, draw a boxplot to represent these data.

b Comment on the distribution of delays. Justify your answer.

The boxplot below summarises the delays at the same airport on another day.

Delay (min)

c Compare the delays to flights arriving at the airport on the two days.

2 The total amount of time a secretary spent on the telephone in a working day was recorded to the nearest minute. The data collected over 40 days are summarised in the table below.

Time (min)	90–139	140–149	150–159	160–169	170–179	180–229
No. of days	8	10	10	4	4	4

Draw a histogram to illustrate these data.

[(c) Edexcel Limited 2003]

S1

3 As part of their job, taxi drivers record the number of miles they travel each day. A random sample of the mileages recorded by taxi drivers Keith and Asif are summarised in the back-to-back stem and leaf diagram below.

Totals	Keith		Asif	Totals
(9)	8 7 7 4 3 2 1 1 0	18	4 4 5 7	(4)
(11)	9 9 8 7 6 5 4 3 3 1 1	19	5 7 8 9 9	(5)
(6)	8 7 4 2 2 0	20	0 2 2 4 4 8	(6)
(6)	9 4 3 1 0 0	21	2 3 5 6 6 7 9	(7)
(4)	6 4 1 1	22	1 1 2 4 5 5 8	(7)
(2)	2 0	23	1 1 3 4 6 6 7 8	(8)
(2)	7 1	24	2 4 8 9	(4)
(1)	9	25	4	(1)
(2)	9 3	26		(0)

Key: 0 | 18 | 4 means 180 miles for Keith and 184 miles for Asif

The quartiles for these two distributions are summarised in the table below.

	Keith	Asif
Lower quartile	191	a
Median	b	218
Upper quartile	221	c

a Find the values of a, b and c.

Outliers are values that lie outside the limits
$Q_1 - 1.5(Q_3 - Q_1)$ and $Q_3 + 1.5(Q_3 - Q_1)$.

b On graph paper, and showing your scale clearly, draw a boxplot to represent Keith's data.

c Comment on the skewness of the two distributions. [(c) Edexcel Limited 2004]

4 The marks, x, for a class of 14 students have $\sum x = 919$ and $\sum x^2 = 60\,773$.

a Calculate the mean and variance of the marks for this class.

A second class of 15 students on the same examination had a mean of 63.8 and standard deviation of 5.58 marks.

b Calculate $\sum y^2$ for the second class.

c Calculate the mean mark of all the students in the two classes.

5 In a particular week, a dentist treats 100 patients. The length of time, to the nearest minute, for each patient's treatment is summarised in the table below.

Time (minutes)	4–7	8	9–10	11	12–16	17–20
Number of patients	12	20	18	22	15	13

Draw a histogram to illustrate these data.

[(c) Edexcel Limited 2003]

6 The following stem and leaf diagram shows the aptitude scores, x, obtained by all the applicants for a particular job.

```
Aptitude score                    3 | 1 means 31

3 | 1 2 9                              (3)
4 | 2 4 6 8 9                          (5)
5 | 1 3 3 5 6 7 9                      (7)
6 | 0 1 3 3 3 5 6 8 8 9                (10)
7 | 1 2 2 4 5 5 5 6 8 8 8 8 9          (14)
8 | 0 1 2 3 5 8 8 9                    (8)
9 | 0 1 2                              (3)
```

a Write down the modal aptitude score.

b Find the three quartiles for these data.

Outliers can be defined to be outside the limits
$Q_1 - 1.0(Q_3 - Q_1)$ and $Q_3 + 1.0(Q_3 - Q_1)$.

c On graph paper, draw a boxplot to represent these data.

For these data, $\sum x = 3363$ and $\sum x^2 = 238\,305$.

d Calculate, to two decimal places, the mean and the standard deviation for these data.

e Use two different methods to show that these data are negatively skewed.

[(c) Edexcel Limited 2004]

7 An airport bus service runs from the city centre, and from the airport, every 10 minutes during the day. The number of passengers on a random sample of journeys from the airport is shown below.

11, 3, 14, 34, 1, 6, 12, 15, 8, 4, 28, 7, 3, 0, 8, 12

Draw a stem and leaf diagram to represent these data.

S1

8 A regular airline passenger keeps a record of the time from when the plane lands until she has collected her bags and exited the terminal. The times, to the nearest minute, for all the journeys she has made in the past six months are summarised in the table below.

Time to exit terminal (minutes)	Number of flights
5–6	12
7–9	15
10–14	12
15–19	8
20–29	11
30–49	14

a Give a reason to support the use of a histogram to represent these data.

b Write down the upper class boundary and the lower class boundary of the class 10–14 minutes.

c On graph paper, draw a histogram to represent these data.

d Calculate estimates of the mean and standard deviation of the time to exit the terminal.

On international flights the passenger had to go through passport control and immigration as well as collecting baggage before exiting the terminal. The shortest time she took to exit on any of the 12 international flights was 25 minutes.

e State, with reasons, what effect excluding the international flights would have on the size of:
  i the mean
  ii the standard deviation of the times to exit the terminal after a flight.

9 The marks on a History examination were 73, 67, 76, 63, 59.

a Calculate the mean and standard deviation of these marks.

For comparisons with other subjects, the marks are to be scaled so that they have a mean of 50 and a standard deviation of 10. This is to be done by using the transformation $Y = \dfrac{(X - a)}{b}$ where X is the original mark and Y is the transformed mark.

b Calculate the values of a and b.

S1

3 Exit ⇒

Summary
Refer to

- Statistical graphs emphasise different aspects of the data or their distribution.
  - Stem-and leaf diagrams give a visual impression of the distribution as well keeping all the detail.
  - Boxplots concentrate on the shape of the distribution, using the quartiles (and sometimes outliers), sacrificing detail for clarity of communication of the shape.
  - Histograms allow you to see the rate of occurrence of observations, which is particularly useful when unequal intervals have been used.   3.1–3.3
- For a data set with n values $(x_1, x_2, \ldots x_n)$ to find the pth percentile, calculate $\left(\frac{p}{100}\right)n$. If it is an integer r then the pth percentile is the midpoint of x_r and x_{r+1}. If it lies between r and $r+1$ then the pth percentile is x_{r+1}.   3.4
- For a set of data, the mean $= \bar{x} = \dfrac{\sum x}{n}$ or $\dfrac{\sum xf}{\sum f}$ and the variance is given by $\sum\dfrac{(x-\bar{x})^2}{n}$ or $\dfrac{\sum(x-\bar{x})^2 f}{\sum f} = \dfrac{\sum x^2}{n} - \bar{x}^2$ or $\dfrac{\sum x^2 f}{\sum f} - \bar{x}^2$ 3.5
- If a set of data values X are related to a set of values Y so that $Y = aX + b$, then
  - Mean of $Y = a \times$ mean of $X + b$
  - Standard deviation of $Y = a \times$ standard deviation of X 3.6
- Comparisons of distributions should make reference to the averages and the spread of the distributions and any skewness which can be seen from graphs or from the values of the summary statistics.   3.7–3.8

Links

In many professional and semi-professional roles extensive use of data is made, so understanding the basic statistical measures and graphical presentation techniques are vital for advancement in many careers.

Biometricians, environmental scientists, forensic scientists, and people working in heath, pharmaceutical, and market research industries all use these techniques extensively.

S1

4

Probability

This chapter will show you how to
- construct sample space diagrams when considering compound outcomes
- draw Venn and tree diagrams to help work out probabilities in more complex contexts
- work with conditional probabilities.

Before you start

You should know how to:

1 Work with fractions.

e.g. Calculate

a $\frac{7}{12} + \frac{1}{3}$ b $\frac{1}{6} \times \frac{1}{2}$

a $\frac{7}{12} + \frac{1}{3} = \frac{7+4}{12} = \frac{11}{12}$

b $\frac{1}{6} \times \frac{1}{2} = \frac{1 \times 1}{6 \times 2} = \frac{1}{12}$

2 Identify the basic outcomes of a simple experiment.

e.g. How many different pairs of letters can be made from the word 'DICE'?

DI   DC   DE
IC   IE   CE

Check in

1 Calculate

a $\frac{1}{4} \times \frac{2}{3}$ b $\frac{1}{4} + \frac{2}{3}$

2 List the possible outcomes when a coin is tossed twice.

S1

A **probability experiment** has outcomes which occur at random.

Imagine an experiment where you roll a die 500 times, and you see 74 'fives'.

The **relative frequency** or **experimental probability** of throwing a five in the experiment is

$$\frac{74}{500} = 0.148$$

However, if the die is fair the **theoretical probability** of throwing a five is

$$\frac{1}{6} = 0.166\ldots$$

Because the outcomes are random, the number of times any particular score appears in an experiment will vary considerably.

The more times you repeat an experiment, the closer the **estimated probability** is likely to be to the theoretical probability.

Out of 500 throws, anything from 65 to 105 fives would not be especially unusual.

An **event** is a set of possible outcomes from an experiment.
So if you throw a die some events could be:
A is the event that a five appears.
B is the event that an even number appears.
C is the event that an odd number appears.

You can use a **Venn diagram** to illustrate events. There is more information about Venn diagrams on page 62.

$A \cup B$ is the **union** of events A and B.
This means A **or** B or both can happen.
2, 4, 5, 6 are the outcomes which satisfy $A \cup B$.

$A \cap C$ is the **intersection** of events A and C.
This means both A **and** C have to happen.
Only the outcome 5 satisfies $A \cap C$.

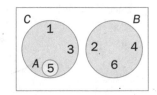

A is inside C

A' means the event 'A does not happen'.
This is the **complementary event**, and $P(A') = 1 - P(A)$.
1, 2, 3, 4, 6 are the outcomes which satisfy A'.

S1

Exercise 4.1

1 A letter is chosen at random from the word CAMBRIDGE.
The events *A*, *B*, *C* and *D* are defined as:

A: A vowel is chosen.
B: The letter B is chosen.
C: A letter in the first half of the alphabet is chosen.
D: A letter is chosen which has only one letter beside it.

a Describe the event *A'* in words.

b For each event *A*, *B*, *C* and *D* write down the outcomes which satisfy it.

c Give the probability of each event *A*, *B*, *C*, *D*.

d List the outcomes which satisfy $A \cap C$.

e Write down $P(A \cap C)$.

f Find $P(A \cap D)$.

g Find $P(A \cup B)$.

Activity

2 **a** Throw a coin 20 times and count the number of times it shows a head.

b Throw the coin another 20 times and count the number of times it shows a head.

c How many times would you 'expect' to see a head in 20 throws?

d Does this happen in both sets of 20 coin throws?

e If you have access to a number of other people's results as well, how often do you see the 'expected number' of heads?

f How many heads (out of 20 throws) are most commonly seen?

3 **a** Throw a die 30 times and count the number of times it shows a five.

b How many times would you 'expect' to see a five in 30 throws?

c Throw the die another 20 times and count the number of times it shows a five.

d How many times would you 'expect' to see a five in 20 throws?

e If you have access to a number of other people's results as well, how often do you see the 'expected number' of fives:
   i in 30 throws?   **ii** in 20 throws?

f How many fives (out of 20 throws) are most commonly seen?

S1

Two events

In contexts where two things happen it is often helpful to construct a table showing the possible outcomes. This is sometimes called a possibility space diagram or a sample space diagram.

Two dice are thrown, and the sum of the scores on the two dice is taken. You can represent this in a two-way table:

Sum	1	2	3	4	5	6
1	2	3	4	5	6	7
2	3	4	5	6	7	8
3	4	5	6	7	8	9
4	5	6	7	8	9	10
5	6	7	8	9	10	11
6	7	8	9	10	11	12

You can use the table to work out the probability of getting a sum of 5.

$$P\{Sum = 5\} = \frac{4}{36}$$

Two dice are thrown, and this time the higher of the scores on the two dice is taken:

High	1	2	3	4	5	6
1	1	2	3	4	5	6
2	2	2	3	4	5	6
3	3	3	3	4	5	6
4	4	4	4	4	5	6
5	5	5	5	5	5	6
6	6	6	6	6	6	6

You can use this table to work out the probability of the higher score being 5.

$$P\{High = 5\} = \frac{9}{36}$$

Exercise 4.2

1 The numbers 1 to 6 are on cards. Two cards are taken at random.

Copy and complete the sample space diagram to show the sum of the numbers on the cards.

Find the probability that the total score is:

a 5   **b** 4   **c** 2

Sum	1	2	3	4	5	6
1			4			
2						
3						
4						
5						
6		8				

2 A coin is tossed and a die is thrown.

a List all the possible outcomes in the sample space.

A head scores 1 and a tail scores 2.

b Construct a sample space diagram to show the total score for this experiment.

In questions 2–5 assume that the dice are fair and six-sided.

3 Two dice are thrown.

a Construct a sample space diagram to show the product of the scores on the two dice.

b Find the probability that the product of scores is:
    i 3   **ii** 5   **iii** 6   **iv** 10

4 Two dice are thrown.

a Construct a sample space diagram to show the lower of the scores on the two dice.

b Find the probability that the lower score is:
    i 3   **ii** 5   **iii** 6

5 Two dice are thrown.

a Copy and complete the sample space diagram to show the difference between the scores on the two dice (the unsigned difference).

b Find the probability that the difference between the scores is:
    i 3   **ii** 5   **iii** 6

Difference	1	2	3
1	0	1	2
2	1	0	1
3	2	1	

S1

Venn diagrams

A **Venn diagram** can be a helpful way to visualise the relationship between events.

The diagram shows each event as a circle within a rectangle.

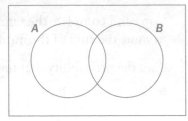

EXAMPLE 1

A school has 97 students in Year 12; 55 students take AS Maths and 32 take AS Chemistry; 31 students take neither AS Maths nor AS Chemistry. How many students take both?

The diagram shows a, b, c and d as the numbers of students in each of those regions.

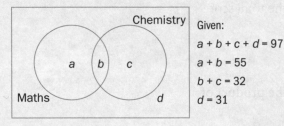

Given:

$a + b + c + d = 97$

$a + b = 55$

$b + c = 32$

$d = 31$

$(a + b) + (b + c) + d = 55 + 32 + 31 = 118 \; [= a + 2b + c + d]$
$a + b + c + d = 97$, so b must be 21. Then $a = 24$ and $c = 11$.
21 students take both AS Maths and AS Chemistry.

The events in the previous example can be represented in terms of probabilities. If you do so, you can see an interesting result.

$P(M) = a + b$ $P(C) = b + c$ $P(M \cap C) = b$

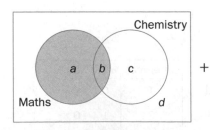   +   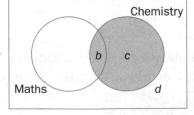   −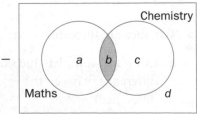

$P(M \cup C) = a + b + c$

=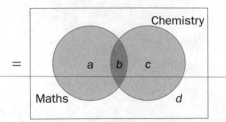

AS

This leads to a general result for two events:

Addition rule: $P(X \cup Y) = P(X) + P(Y) - P(X \cap Y)$

In the special case where $P(X \cap Y) = 0$ (X and Y are **mutually exclusive**), the probability of X or Y happening is just the sum of the probabilities of X and of Y.

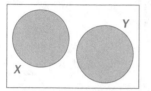

EXAMPLE 2

For two events A and B,
$$P(A) = 0.7, \quad P(B) = 0.4, \quad P(A \cap B) = 0.3$$

Find

i $P(A \cup B)$ **ii** $P(A' \cap B')$

iii $P(A' \cap B)$ **iv** $P(A' \cup B)$

Draw a Venn diagram and label the four regions:

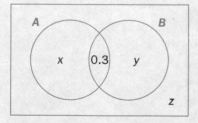

$(A \cap B) = 0.3$ so $x = 0.4$ from $P(A)$ and $y = 0.1$ from $P(B)$.
$0.3 + 0.4 + 0.1 = 0.8$
so $z = 0.2$

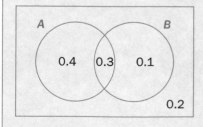

EXAMPLE 2 (CONT.)

Now find the required probabilities:

i $P(A \cup B) = 0.8$

ii $P(A' \cap B') = 0.2$

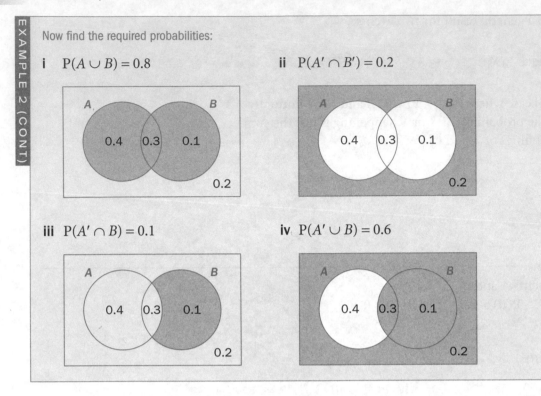

iii $P(A' \cap B) = 0.1$

iv $P(A' \cup B) = 0.6$

You sometimes need to extract the necessary information from the context described.

EXAMPLE 3

Two-thirds of the pupils in a class have a mobile phone. Half of the pupils have an MP3 player. A quarter of the pupils have neither.
What proportion of the class have both?

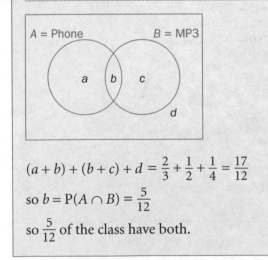

$(a + b) + (b + c) + d = \dfrac{2}{3} + \dfrac{1}{2} + \dfrac{1}{4} = \dfrac{17}{12}$

so $b = P(A \cap B) = \dfrac{5}{12}$

so $\dfrac{5}{12}$ of the class have both.

Three events can be shown as three intersecting circles, giving eight possible regions.

Venn diagrams will not do the general case for four events – special cases can be drawn, but you are not going to meet them at this level.

EXAMPLE 4

$P(A) = 0.6$, $P(B) = 0.5$, $P(C) = 0.3$, $P(A \cap B) = 0.3$
$P(A \cap C) = 0.1$, $P(A \cap B \cap C) = 0.1$, $P(A' \cap B' \cap C') = 0.1$

Draw and complete a Venn diagram for the three events A, B and C.

A good start is to draw blank circles, and then allocate letters to each region.

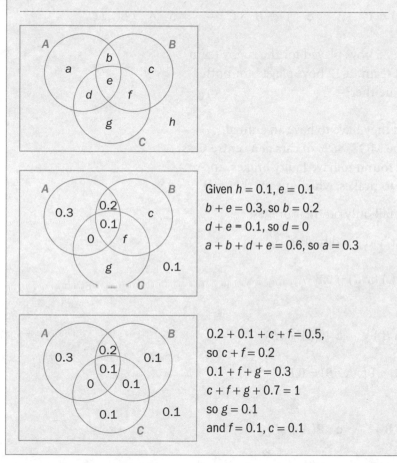

Given $h = 0.1$, $e = 0.1$
$b + e = 0.3$, so $b = 0.2$
$d + e = 0.1$, so $d = 0$
$a + b + d + e = 0.6$, so $a = 0.3$

$0.2 + 0.1 + c + f = 0.5$,
so $c + f = 0.2$
$0.1 + f + g = 0.3$
$c + f + g + 0.7 = 1$
so $g = 0.1$
and $f = 0.1$, $c = 0.1$

Exercise 4.3

1 For each of the following, draw a copy of the diagram below and shade in the area representing the set.

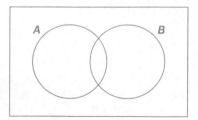

  a $A \cap B$ **b** $A \cup B$ **c** $A' \cap B$ **d** $A \cup B'$ **e** $A' \cap B'$

2 For each of the following, draw a copy of the diagram below and shade in the area representing the set.

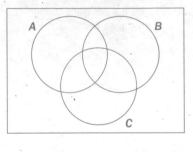

a $A \cap C$ **b** $(A \cup B) \cap C$ **c** $A' \cap B \cap C$ **d** $A \cup B' \cup C$

3 There are 80 boys in Year 10; 26 boys played for the rugby team and 17 played for the cricket team. If 12 boys played for both teams, how many played for neither?

4 Once cars are three years old they have to have an annual roadworthiness test called the MOT. 86% of cars at a centre pass the MOT. 9% of the cars are found to have faulty brakes, and 11% have a fault not related to brakes, which means they fail.

a What proportion of cars fail only on their brakes?

b What proportion fail, but had good brakes?

5 $P(A) = 0.6$, $P(B) = 0.5$, $P(A \cup B) = 0.8$

Calculate

a $P(A \cap B)$ **b** $P(A' \cap B')$ **c** $P(A' \cap B)$

6 $P(A) = 0.7$, $P(A \cap B) = 0.5$, $P(A \cup B) = 0.8$

Calculate

a $P(B)$ **b** $P(A' \cap B')$ **c** $P(A' \cap B)$

7 $P(A) = \frac{2}{3}$, $P(B) = \frac{1}{4}$, $P(A \cup B) = \frac{3}{4}$

Calculate

a $P(A \cap B)$ **b** $P(A \cap B')$ **c** $P(A' \cap B)$

8 $P(A) = \frac{1}{2}$, $P(A \cap B) = \frac{1}{3}$, $P(A \cup B) = \frac{3}{4}$

Calculate

a $P(B)$ **b** $P(A \cap B')$ **c** $P(A' \cap B)$

9 $P(A') = 0.6$, $P(A' \cap B') = 0.4$, $P(A \cap B') = 0.1$

Calculate

a $P(B)$ b $P(A \cap B)$ c $P(A \cup B)$

10 A survey of a primary school class found that one-third of the pupils had a pet dog, a quarter had a pet hamster and one-sixth of the class had both.
A child is chosen at random from the class.
What is the probability that the child has:

a at least one of a pet dog or hamster?

b a pet dog but not a hamster?

11 $P(A) = 0.5$, $P(B) = 0.5$, $P(C) = 0.2$
$P(A \cap B) = 0.2$, $P(B \cap C) = 0$, $P(A' \cap B' \cap C') = 0.1$

Calculate

a $P(A \cap C)$

b $P(A \cap B' \cap C')$

12 A Year 12 group has 112 pupils; 42 take French, 65 take Maths and 32 take Physics. Everyone who takes Physics also takes Maths, but no one takes all three subjects. 12 pupils take Maths and French. How many pupils in the group take none of Maths, French and Physics?

13 A lecturer does an informal survey of his first-year undergraduate politics students. Half the students claim to read the *Daily Telegraph*, half claim to read *The Times* and half claim to read the *Guardian*; $\frac{1}{12}$ say they read none of these.

No one claims to read all three papers; $\frac{1}{6}$ claim to read both the *Daily Telegraph* and *The Times*, and $\frac{1}{6}$ claim to read both the *Daily Telegraph* and the *Guardian*.

a How many claim to read both *The Times* and the *Guardian*?

b How many claim to read only the *Guardian*?

c How many claim to read only the *Daily Telegraph*?

S1

Sampling without replacement

Consider a bag which has three red and five blue beads in it. If you take a bead out at random, and then take out another without replacing the first, you can represent the possible outcomes in a **probability tree** diagram.

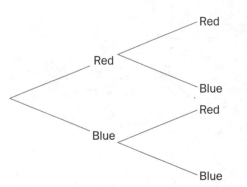

You can put probabilities on the branches to complete the diagram.

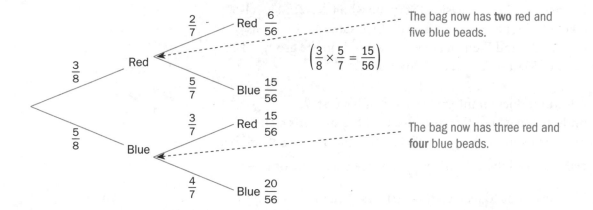

The bag now has **two** red and five blue beads.

$$\left(\frac{3}{8} \times \frac{5}{7} = \frac{15}{56}\right)$$

The bag now has three red and **four** blue beads.

You can now scan along the branches to identify any possible outcome. For example,

P (both beads the same colour) = P (both red) + P (both blue)

$$= \left(\frac{3}{8} \times \frac{2}{7}\right) + \left(\frac{5}{8} \times \frac{4}{7}\right)$$

$$= \frac{6}{56} + \frac{20}{56}$$

$$= \frac{26}{56}$$

$$= \frac{13}{28}$$

Think × for **and**, + for **or**:
red and red   or   blue and blue
= red × red   +   blue × blue

S1

Sampling with replacement

If the first bead was returned to the bag before
the second one was selected, the probabilities
of red or blue would still be $\frac{3}{8}$ and $\frac{5}{8}$ at the second stage:

Now, P(both beads the same colour) = P (both red) + P (both blue)

$$= \frac{9}{64} + \frac{25}{64}$$

$$= \frac{34}{64}$$

$$= \frac{17}{32}$$

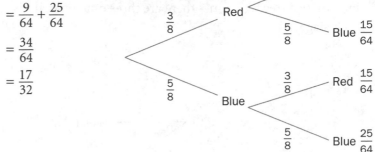

EXAMPLE 1

A disease is known to affect 1 in 10 000 people. It can be fatal,
but it is treatable if it is detected early.
A screening test for the disease shows a positive result for
99% of people with the disease.
The test shows positive for 1% of people who do not have
the disease.
For a population of 1 million people:

A screening test will usually
detect the presence of bodies
which are almost always
present with the disease, but
which also occur naturally in a
small proportion of people.

a how many would you expect to have the disease and test
   positive?

b how many would you expect to test positive?

First draw a tree diagram:

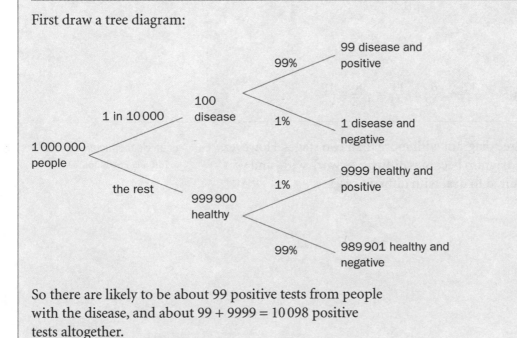

So there are likely to be about 99 positive tests from people
with the disease, and about 99 + 9999 = 10 098 positive
tests altogether.

More complex tree diagrams

EXAMPLE 2

There are four red, three green and five blue discs in a bag. Find the probability that two discs drawn without replacement are the same colour.

First draw a tree diagram – there are three outcomes at each stage.

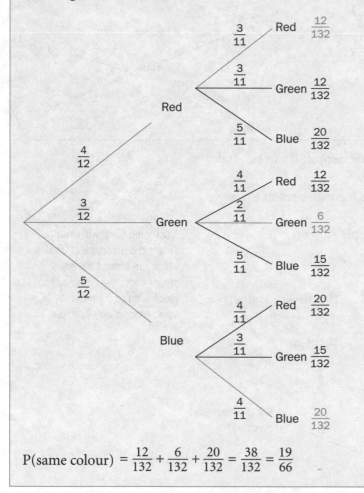

It is much easier to keep all the denominators the same in each stage.

$$P(\text{same colour}) = \frac{12}{132} + \frac{6}{132} + \frac{20}{132} = \frac{38}{132} = \frac{19}{66}$$

You can use a tree diagram with more than two stages. However, in practice the diagram becomes difficult to work with, and you will not be required to deal with difficult cases.

A neat, clear structure to your diagrams is essential.

EXAMPLE 3

A bag has three red beads and five blue beads.
A bead is taken from the bag, its colour is noted, and it is returned to the bag. This is done three times. Find the probability that all three beads are the same colour.

Find draw a tree diagram.

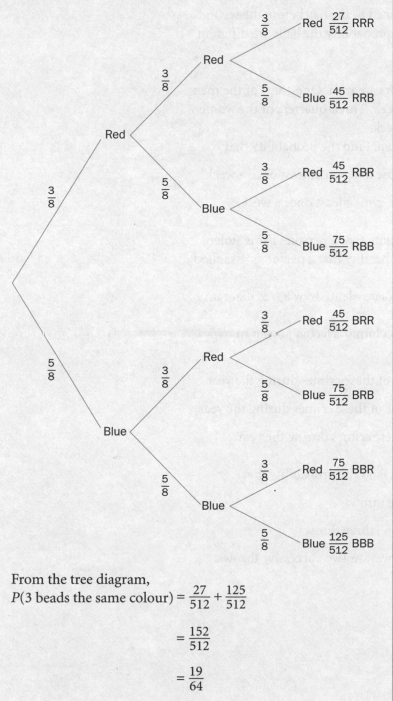

It helps to summarize the outcomes at the ends of the branches eg RRR, RRB.

From the tree diagram,
$$P(\text{3 beads the same colour}) = \frac{27}{512} + \frac{125}{512}$$

$$= \frac{152}{512}$$

$$= \frac{19}{64}$$

S1

Exercise 4.4

1 A bag contains five blue and three green balls. A ball is chosen at random, its colour noted, and the ball returned to the bag. A second ball is chosen.

a Find the probability that the two balls are different colours.

b If the first ball is not returned to the bag before the second ball is chosen, what is the probability the balls are different colours?

2 At a gym, 60% of the members are men. One-third of the men use the gym at least once a week. Three-quarters of the women use the gym at least once a week.
A member is chosen at random. Find the probability that

a it is a man who does not use the gym at least once a week

b it is a person who uses the gym at least once a week.

3 In a certain town, the probability that a person's car is stolen during a year is 0.06. The probability that a person is assaulted is 0.03.
Assuming these events are independent, draw a tree diagram to represent this information.
Find the probability that a randomly selected person in the town

a is not the victim of either of these crimes during the year

b is the victim of exactly one of these crimes during the year

c is the victim of both of these crimes during the year

4 A coin is thrown three times. Find the probability that

a it shows heads on all three throws

b it shows the same face on all three throws

c it does not land the same way on two successive throws.

5 Bag A contains five blue and three green balls. A ball is chosen at random, the colour is noted and it is **not** returned to the bag. A second ball is chosen.

   a Find the probability that the two balls are the same colour.

   Bag B contains 50 blue and 30 green balls. Again, a ball is chosen at random, the colour is noted and it is **not** returned to the bag before a second ball is chosen.

   b Find the probability that the two balls are the same colour.

6 A bag contains four blue, four red and four green balls. Two balls are removed at random, one at a time, and without replacement. Find the probability that:

   a the second ball drawn is a red

   b both balls are blue

   c neither ball is green

   d at least one ball is green.

7 A bag contains ten counters: four white, three green and three red. Counters are removed at random, one at a time, and without replacement. Find the probability that:

   a the first counter is red

   b the first three counters are all white

   c the first three counters are all different colours.

In the previous section you considered a screening test for a rare disease. Even though the test was remarkably accurate, less than one in 100 of the positive results come from someone with the disease.

Of 10 098 (99 + 9999) positive results 9999 were from healthy people, so the **conditional probability** that somebody is healthy **given that** they have a positive result is

$$P(\text{Healthy} \mid \text{positive test result}) = \frac{9999}{10098} = 0.9902$$

In situations like this it is important that the patient is not told that they have the disease on the basis of a positive result from a screening test.

> The conditional probability of an event A occurring given that an event B has already occurred can be written as $P(A \mid B)$.

EXAMPLE 1

A GP practice encourages elderly people to have a flu vaccination each year. The doctors say that the vaccination reduces the likelihood of having flu from 40% to 10%. If 45% of the elderly people in the practice have the vaccination, find the probability that an elderly person chosen at random from the practice:

a gets flu

b had the vaccination, given that they get flu.

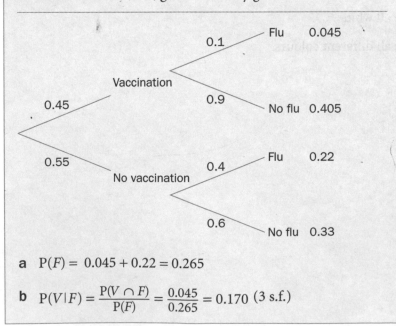

a $P(F) = 0.045 + 0.22 = 0.265$

b $P(V \mid F) = \dfrac{P(V \cap F)}{P(F)} = \dfrac{0.045}{0.265} = 0.170$ (3 s.f.)

Note:

$P(V \mid F) = \dfrac{P(V \cap F)}{P(F)}$

not $\dfrac{P(V)}{P(F)}$

S1

Venn diagrams can be useful when looking at conditional probability. Consider two events *A* and *B*, with the probabilities shown in the Venn diagram.

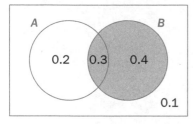

P(*A* | *B*) is the probability of *A* given that *B* has occurred.

So $P(A|B) = \frac{0.3}{0.7} = \frac{3}{7}$

> Generally, for two events *A* and *B*
>
> $P(A|B) = \frac{P(A \cap B)}{P(B)}$

This is often known as the **multiplication law** of probability.

Note that generally $P(A|B) \neq P(B|A)$.

In the diagram above

$P(B|A) = \frac{P(B \cap A)}{P(A)} = \frac{0.3}{0.5} = \frac{3}{5}$

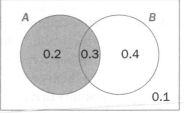

EXAMPLE 2

$P(A) = 0.75$, $P(B) = 0.35$, $P(A \cup B) = 0.9$

Find **a** $P(A \cap B)$ **b** $P(A|B)$ **c** $P(B|A')$

---

a $P(A \cup B) = P(A) + P(B) - P(A \cap B)$
so $P(A \cap B) = 0.75 + 0.35 - 0.9 = 0.2$

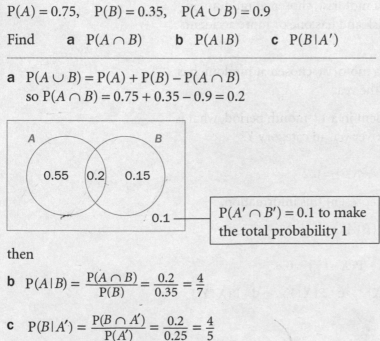

$P(A' \cap B') = 0.1$ to make the total probability 1

then

b $P(A|B) = \frac{P(A \cap B)}{P(B)} = \frac{0.2}{0.35} = \frac{4}{7}$

c $P(B|A') = \frac{P(B \cap A')}{P(A')} = \frac{0.2}{0.25} = \frac{4}{5}$

Exercise 4.5

1 95% of drivers wear seat belts. 60% of car drivers involved in serious accidents die if they are not wearing a seat belt, whereas 80% of those who do wear a seat belt survive.

   a Draw a tree diagram to show this information.

   b What is the probability that a driver in a serious accident did not wear a seat belt and survived?

2 At an electrical retailer's, one-third of the light bulbs are from company X, and the rest from Company Y. A report shows that 3% of light bulbs from Company X are faulty, and that 2% from Company Y are faulty.

   a If the retailer chooses a bulb at random from stock and tests it, what is the probability that it is faulty?

   b If the bulb is faulty, what is the probability that it came from Company Y?

3 An insurance company classifies drivers in three categories. X is 'low risk', and they represent 20% of drivers who are insured. Y is 'moderate risk' and they represent 70% of the drivers. Z is 'high risk'.
The probability that a category X driver has one or more accidents in a 12-month period is 2%, and the corresponding probabilities for Y and Z are 5% and 9%.

   a Find the probability that a motorist, chosen at random, is assessed as a category Y risk and has one or more accidents in the year.

   b Find the probability that a motorist, chosen at random, has one or more accidents in the year.

   c If a customer has an accident in a 12-month period, what is the probability that the driver was in category Y?

4 $P(A) = 0.6$, $P(B) = 0.5$, $P(A \cap B) = 0.2$

   a Draw a Venn diagram to represent this information.

   b Find   **i** $P(A|B)$ **ii** $P(B|A)$

5 $P(X) = 0.6$, $P(X \cap Y) = 0.3$, $P(X \cup Y) = 0.8$

   Find   **a** $P(Y)$ **b** $P(Y|X)$ **c** $P(X|Y)$ **d** $P(X|Y')$

6   In a school there are 542 pupils, 282 of whom are girls.
    364 pupils walk to school, of whom 153 are girls. Find the
    probability that a pupil chosen at random:

   a   is a boy

   b   is a boy who does not walk to school

   c   does not walk to school given that the pupil is a boy

   d   is a girl, given that the pupil walks to school.

7   There are 173 pupils in Year 13 in a school. There are 25
    prefects in Year 13, of whom 7 are House Captains.

    Find the probability that a Year 13 pupil chosen at random:

   a   is a prefect

   b   is a House Captain, given that the pupil is a prefect.

8   Of the employees in a large factory one-sixth travel to work by
    bus, one-third by train, and the rest by car. Those travelling
    by bus have a probability of $\frac{1}{4}$ of being late, those

    by train will be late with probability $\frac{1}{5}$ and those by car will be

    late with probability $\frac{1}{10}$.

    Draw and complete a tree diagram and calculate the probability
    that an employee chosen at random will be late.

9   The homework diaries and homework of two pupils are
    examined. There is a probability of 0.4 that A does not write
    in the given homework correctly. She always does the homework
    if she writes it in, but never checks if it is not written in.
    There is a probability of 0.8 that B writes in the homework
    correctly, and when she does she will do it 90% of the time;
    if she has nothing written in then she checks with a friend
    who knows the homework 50% of the time.
    She does the homework if she is given it by the friend.
    Assume that A and B act independently.
    Draw a tree diagram representing this information.

   a   Find the probability that pupil A does her homework on
       a particular night.

   b   Find the probability that both pupils do their homework on
       a particular night.

   c   If a homework was checked and was not done, find the
       probability that it was pupil A.

Independence

Two events A and B are **independent** if the outcome of A does not affect the outcome of B, and vice versa.

So $P(A|B) = P(A)$ ⟺ A and B are independent.

The probability of A occurring given that B has already occurred will just be the probability of A.

EXAMPLE 1

A set of 40 cards shows a number from 1 to 10 of one of four geometrical symbols.
Circles and squares are shown in grey, rectangles and ellipses are blue.
The pack of cards is shuffled and the top card is turned over.
Let C be the event 'card shows a circle', F be the event 'card shows 5 symbols' and G be the event 'the card is grey'.

a Show that C and F are independent events.

b Show that C and G are not independent events.

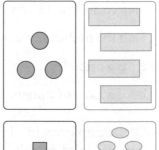

a $P(C) = \frac{10}{40}$ (there are 10 circle cards)

$P(F) = \frac{4}{40}$ (there are 4 cards showing 5 symbols)

$P(C \cap F) = \frac{1}{40}$ (there is only one card with 5 circles)

$P(C|F) = \frac{P(C \cap F)}{P(F)} = \frac{\frac{1}{40}}{\frac{4}{40}} = \frac{1}{4} = P(C)$

Since $P(C|F) = P(C)$, C and F are independent events.

b $P(G|C) = 1$, since if you know the card has circles then you know it is grey

$P(G) = 0.5$
So $P(G|C) \neq P(G)$, and the events G and C are not independent.

Knowing that F has happened has given no additional information concerning whether C is likely to happen or not.

You can test for independence by comparing $P(X|Y)$ with $P(X)$, or equivalently you compare $P(X \cap Y)$ with $P(X) \times P(Y)$.

Mutually exclusive events

Two events A and B are **mutually exclusive** if they cannot occur at the same time.

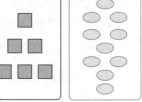

For mutually exclusive events, $P(A \cup B) = P(A) + P(B)$.

The general relationship is:
$P(A \cup B) = P(A) + P(B) - P(A \cap B)$

Exhaustive events

A set of events is **exhaustive** if it covers all possible outcomes.

EXAMPLE 2

Two fair dice are thrown. From the events which are listed below, give two which are:

a mutually exclusive

b exhaustive

c not independent.

A: The two dice show the same number.
B: The sum of the two scores is at least 5.
C: At least one of the two numbers is a 5 or a 6.
D: The sum of the two scores is odd.
E: The largest number shown is a 6.
F: The sum of the two scores is less than 8.

a A and D are mutually exclusive because if the two dice show the same number, the sum has to be an even number.

b B and F are exhaustive because the only outcomes not in B are that the sum is 2, 3 or 4 and these are all in F.

c A and D are not independent (because they are mutually exclusive)
C and E are not independent since $P(C|E) = 1$
(if E happens, then you know C must happen).

There are other possibilities for part c.

Here are some other terms that can be helpful.

Partition: a group of sets which are exhaustive and mutually exclusive form a **partition**. The whole outcome space has been split into disjoint events, so their probabilities total 1, and there is no overlap between any pair.
Compound events can be evaluated simply by going through the group, and seeing whether each set is to be included.

You will not be assessed on knowledge of the term 'partition'.

Complementary event: this is a two-event partition. If A and B are complementary then $P(B) = 1 - P(A)$. The simplest way of identifying a complementary pair is A and 'not A' (A').

EXAMPLE 3

Over the course of a season, a hockey team play 40 matches, in different conditions, with the following results.

This is a **two-way table**.

		Weather		
		Good	Bad	Total
Result	Win	13	6	19
	Draw	5	3	8
	Lose	7	6	13
	Total	25	15	40

For a match chosen at random from the season:
G is the event 'Good weather'
W is the event 'Team wins'
D is the event 'Team draws'
L is the event 'Team loses'.

a Find the probabilities:
   i $P(G)$
   ii $P(G \cap D)$
   iii $P(D \mid G)$

b Are the events D and G independent?

a i There are 25 good weather matches out of the 40 so
$$P(G) = \frac{25}{40} = \frac{5}{8}$$

ii $G \cap D$ is a draw in good weather, and there are five of those so $P(G \cap D) = \frac{5}{40} = \frac{1}{8}$

iii $P(D \mid G) = \frac{P(G \cap D)}{P(G)} = \frac{\frac{1}{8}}{\frac{5}{8}} = \frac{1}{5}$

b $P(D) = \frac{8}{40} = \frac{1}{5}$, so since $P(D \mid G) = P(D)$ the events D and G are independent.

Exercise 4.6

1 A and B are independent events. $P(A) = 0.7$, $P(B) = 0.4$.
   Find: **a** $P(A \cap B)$ **b** $P(A \cup B)$ **c** $P(A' \cap B)$

2 $P(A) = 0.7$, $P(B) = 0.4$, $P(A \cup B) = 0.82$.
   Show that A and B are independent.

3 $P(A) = 0.5$, $P(B \mid A) = 0.6$, $P(B') = 0.7$
   Show that A' and B are mutually exclusive.

S1

4 X and Y are independent events with $P(X) = 0.4$ and $P(Y) = 0,5$.

   a Write down $P(X \mid Y)$.

   b Write down $P(Y \mid X)$.

   c Calculate $P(X' \cap Y)$.

5 The results of a traffic survey of the colour and type of car are given in the following table:

	Saloon	Hatchback
Silver	65	59
Black	27	22
Other	16	19

   One car is selected from the group at random. Find the probability that the selected car is:

   i   a silver hatchback

   ii a hatchback

   iii a hatchback, given that it is silver.

   Show that the type of car is not independent of its colour.

6 Consider the following possible events when a blue and a white die are rolled:

   A: the total is 2 B: the white is a multiple of 2
   C: the total is < 10 D: the white is a multiple of 3
   E: the total is > 7 F: the total is > 9

   Which of the following pairs are exhaustive and which are mutually exclusive?

   a A, B **b** A, D **c** C, E

   d C, F **e** B, D **f** A, E

7 Four unbiased coins are tossed together. For the events A to D below, say whether the statements **a** to **d** are true or false, and give a reason for each answer. (X' means NOT X)

   A: no heads B: at least one head
   C: no tails D: at least two tails

   a A and B are mutually exclusive

   b A and B are exhaustive

   c B and D are exhaustive

   d A' and C' are mutually exclusive

S1

1 A fair die has six faces numbered 1, 2, 2, 3, 3 and 3. The die is rolled twice and the number showing on the uppermost face is recorded each time.

Find the probability that the sum of the two numbers recorded is at least 5.

[(c) Edexcel Limited 2004]

2 A bag contains eight purple balls and two pink balls. A ball is selected at random from the bag and its colour is recorded. The ball is not replaced. A second ball is selected at random and its colour is recorded.

a Draw a tree diagram to represent the information.

Find the probability that

b the second ball selected is purple

c both balls selected are purple, given that the second ball selected is purple.

3 For the events A and B,

$P(A) = 0.5$, $P(A' \cap B) = 0.27$ and $P(A' \cup B) = 0.53$.

a Draw a Venn diagram to illustrate the complete sample space for the events A and B.

b Write down the value of $P(B)$.

c Find $P(B|A)$.

d Determine whether or not A and B are independent.

4 The events A and B are such that $P(A) = \frac{5}{12}$, $P(B) = \frac{2}{3}$ and $P(A' \cap B') = \frac{1}{12}$.

a Find:

  i $P(A \cap B')$
  ii $P(A|B)$
  iii $P(B|A)$.

b State, with a reason, whether or not A and B are:

  i mutually exclusive
  ii independent.

5 Two events A and B are mutually exclusive. $P(A) = \frac{1}{2}$, $P(B) = \frac{1}{3}$

  a Find $P(A|B)$.

  b Find $P(A \cup B)$.

  c Are events A and B independent? You must provide a reason.

6 Walker's disease is a rare tropical disease, which is known to be present in only 0.1% of the population. A new screening test has been analysed and shows a 98% probability of showing positive when the person tested has the disease and only 0.2% of showing positive when the person does not have the disease. A person is selected at random from the population and given the screening test.

  a What is the probability that the test will show positive?

  b What is the probability that the person does not have the disease, given that the test showed positive?

  c Jane is a doctor who is unhappy with guidelines which say that patients should be told immediately if the test shows positive.
  Explain how she could use the answer to part **b** to argue that these guidelines are not appropriate.

7 A computer-based testing system gives the user a hard question if they got the previous question correct and an easy question if they got previous question wrong. The first question is randomly chosen to be hard or easy.

The probability of Benni getting an easy question right is $\frac{2}{3}$,

and the probability he gets a hard question right is $\frac{1}{4}$.

  a Draw a tree diagram to represent what can happen in the first two questions Benni has in a test.

  b Find the probability that Benni gets his first two questions correct.

  c Find the probability that the first question was hard, given that Benni got both of his first two questions correct.

S1

8 For the events A and B:

    a   explain in words the meaning of the term $P(B|A)$,

    b   sketch a Venn diagram to illustrate the relationship $P(B|A) = 0$.

Three companies operate a bus service along a busy main road. Amber buses run 50% of the service and 2% of their buses are more than 5 minutes late. Blunder buses run 30% of the service and 10% of their buses are more than 5 minutes late. Clipper buses run the remainder of the service and only 1% of their buses run more than 5 minutes late.

Jean is waiting for a bus on the main road.

    c   Find the probability that the first bus to arrive is an Amber bus that is more than 5 minutes late.

Let A, B and C denote the events that Jean catches an Amber bus, a Blunder bus and a Clipper bus respectively. Let L denote the event that Jean catches a bus that is more than 5 minutes late.

    d   Draw a Venn diagram to represent the events A, B, C and L. Calculate the probabilities associated with each region and write them in the appropriate places on the Venn diagram.

    e   Find the probability that Jean catches a bus that is more than 5 minutes late.

[(c) Edexcel Limited 2002]

9 A car dealer offers purchasers a three-year warranty on a new car.

He sells two models, the Zippy and the Nifty. For the first 50 cars sold of each model the number of claims under the warranty is shown in the table.

	Claim	No claim
Zippy	35	15
Nifty	40	10

One of the purchasers is chosen at random. Let A be the event that no claim is made by the purchaser under the warranty and B the event that the car purchased is a Nifty.

    a   Find $P(A \cap B)$.

    b   Find $P(A')$.

Given that the purchaser chosen does not make a claim under the warranty:

    c   find the probability that the car purchased is a Zippy

    d   show that making a claim is not independent of the make of the car purchased.
Comment on this result.

[(c) Edexcel Limited 2003]

10 In a factory, machines X, Y and Z are all producing metal rods of the same length. Machine X produces 25% of the rods, machine Y produces 45% and the rest are produced by machine Z. Of their production of rods, machines X, Y and Z produce 4%, 5% and 2% defective rods respectively.

    a Draw a tree diagram to represent this information.

    b Find the probability that a randomly selected rod is

       i   produced by machine Y and is not defective

       ii is not defective.

    c Given that a randomly selected rod is not defective, find the probability that it was produced by machine Y.

11 A golfer enters two tournaments. He estimates the probability that he wins the first tournament is 0.6, that he wins the second tournament is 0.4 and that he wins them both is 0.35.

    a Find the probability that he does not win either tournament.

    b Show, by calculation, that winning the first tournament and winning the second tournament are not independent events.

    c The tournaments are played in successive weeks. Explain why it would be surprising if these were independent events.

12 The events A and B are independent such that $P(A) = \frac{1}{2}$ and $P(B) = \frac{1}{3}$.

    Find:

    a $P(A \cap B)$

    b $P(A' \cap B')$

    c $P(A|B)$.

S1

13 A fair die has six faces numbered 4, 4, 4, 5, 6 and 6. The die is rolled twice and the number showing on the uppermost face is recorded each time.

Find the probability that the sum of the two numbers recorded is at least 10.

14 Events A and B are defined in the sample space S. The events A and B are independent.

Given that $P(A) = 0.3$, $P(B) = 0.4$ and $P(A \cup B) = 0.65$, find:

a $P(B)$

b $P(A|B)$

A and C are mutually exclusive and $P(C) = 0.5$.

c Find $P(A \cup C)$.

15 The events A and B are such that $P(A) = \frac{1}{3}$, $P(B) = \frac{2}{3}$ and $P(A \cap B) = \frac{1}{4}$.

a Represent these probabilities in a Venn diagram.

Hence, or otherwise, find

b $P(A'|B')$

c $P(B|A)$.

16 a If A and B are two events which are statistically independent, write down expressions for $P(A \cap B)$ and $P(A \cup B)$ in terms of $P(A)$ and $P(B)$.

b Anji and Katrina are keen cinema goers, but they decide each Friday independently of one another whether they go to the cinema. On any given Friday, the probability of both going to the cinema is $\frac{1}{3}$, and the probability that at least one of them goes is $\frac{5}{6}$.

Find the possible values for the probability that Anji goes to the cinema on a Friday.

(8)

17 Of the pupils who took English in a certain school one year, 60% of them took History, 30% of them took Religious Studies and 10% took both History and Religious Studies.
One of the pupils taking English is chosen at random.

    a Find the probability that this pupil took neither History nor Religious Studies.

    b Given that the pupil took exactly one of History and Religious Studies, find the probability it was History.

18 Two identical bags each contain 12 discs, which are identical except for colour. One bag (*A*) contains six red and six blue discs, and the other (*B*) contains eight red and four blue discs.

    a A bag is selected at random and a disc is selected from it. Draw a tree diagram, illustrating this situation, and calculate the probability that the disc drawn will be red.

    b The disc selected is now returned to the same bag, along with another two of the same colour, and another disc is chosen from that bag. Find the probability that:

       i it is the same colour as the first disc drawn
       ii bag *A* was used, given that two discs of the same colour have been chosen.

Summary

- The relative frequency of an event happening can be used as an estimate of the probability of that event happening. The estimate is more likely to be close to the true probability if the experiment has been carried out a large number of times.

- A two-way table can be used to show the possible outcomes of a compound event such as throwing two dice.

- Venn diagrams are useful when you have information about single events and also their union or intersection.

- Tree diagrams are useful when you know the probabilities of each stage of compound events. You multiply along the branches to get the probability of a pathway, and the probabilities of different pathways can be added.

- The conditional probability of V given F is $P(V|F) = \dfrac{P(V \cap F)}{P(F)}$.

  Be careful to work out the probability of both V and F happening directly and not from the probabilities of V and F happening individually.

- Events V and F are independent if $P(V|F) = P(V)$ i.e. knowing that F has happened has given no information about the likelihood of V happening.

Links

Conditional probability reasoning is a fundamental component of using DNA evidence in trials.

Evaluating risk is a fundamental part of our everyday lives – and it is mostly done very informally, so having a good understanding of the way likelihoods of different events combine can help you to make better informed judgements.

5

Correlation

This chapter will show you how to
- calculate the correlation coefficient from raw bivariate data and from summary statistics
- interpret the value of the correlation coefficient
- code bivariate data and calculate the correlation coefficient from coded data.

Before you start

You should know how to:

1 Draw a scatter graph.

2 Calculate simple summations such as $\sum x$, $\sum xy$.

e.g. The table gives test results for five pupils. Calculate $\sum x$ and $\sum xy$ for these marks.

English mark (x)	1	8	15	18	23
Arithmetic mark (y)	3	14	8	20	19

$\sum x = 1 + 8 + 15 + 18 + 23 = 65$

$\sum xy = (1 \times 3) + (8 \times 14) + (15 \times 8)$
$\quad\quad + (18 \times 20) + (23 \times 19)$
$\quad\quad = 3 + 112 + 120 + 360 + 437 = 1032$

Check in

1 For the data set shown, plot a scatter graph.

x	5	8	9	12	14
y	8	9	9	13	18

2 For the data set in question 1, calculate $\sum x$, $\sum xy$.

Here are two scatter diagrams, or scatter graphs, relating height to handspan and height to IQ for two sets of people.

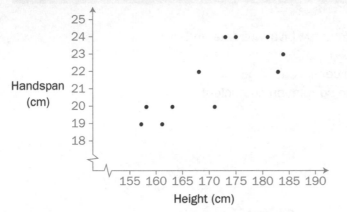

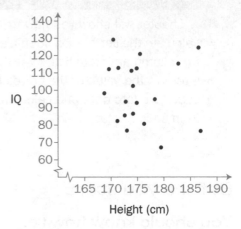

Taller people tend to have a larger handspan than shorter people.

There does not seem to any tendency for taller people to have either higher or lower IQs than shorter people.

Correlation describes the strength of the relationship between two variables.
Paired data is often known as **bivariate data**.

> The **product moment correlation coefficient, r**, is a measure of how well the data fit a straight line. It is a numerical value which lies between −1 and +1.

Here are some examples of scatter graphs with different values of r.

Weight of premature babies in a children's hospital

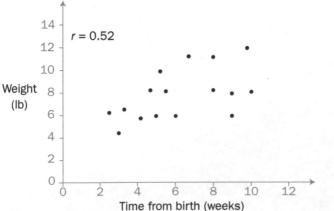

Cost of fuel at a petrol station

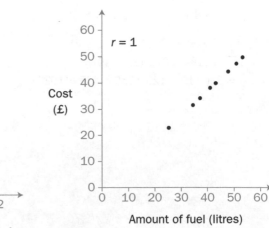

Weak positive correlation – babies should put on weight as they get older, but the correlation is not very strong because there are many other factors involved.

Perfect positive correlation – the cost is directly proportional to the amount of fuel.

Age and the maximum distance at which a sample of drivers could read a motorway sign

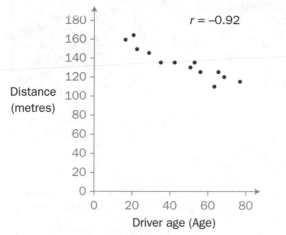

As part of a forensic laboratory's standardising programme two of the technicians are each given a number of matching samples for analysis.

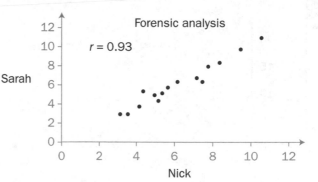

Strong negative correlation – generally, people's eyesight gets worse as they get older.

Positive strong correlation – high values occur together and low values also occur together. Since Nick and Sarah were measuring the same quantity in each case, the lab would ideally want their measurements to lie exactly on the straight line $y = x$. Here, the lie quite close to it, so the technicians are reasonably accurate.

This graph shows the temperature of some cakes at various times after they have been taken out of the oven.

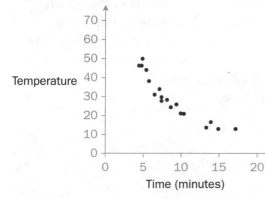

The number of churches and the number of police officers in six towns and cities

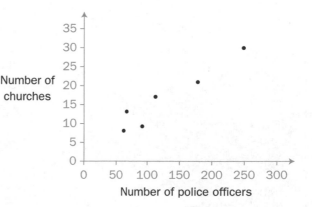

The product moment correlation coefficient is a measure of the degree of linear association. Here the temperature seems to be reasonably predictable, but the relationship is not linear and you should not calculate the pmcc for data like this.

Spurious (positive) correlation – the relationship is through a hidden third variable (the size of the town).

You can get a rough idea of the type of correlation by dividing the
scatter graph into quadrants.

EXAMPLE 1

For each of the scatter diagrams state the type of correlation.

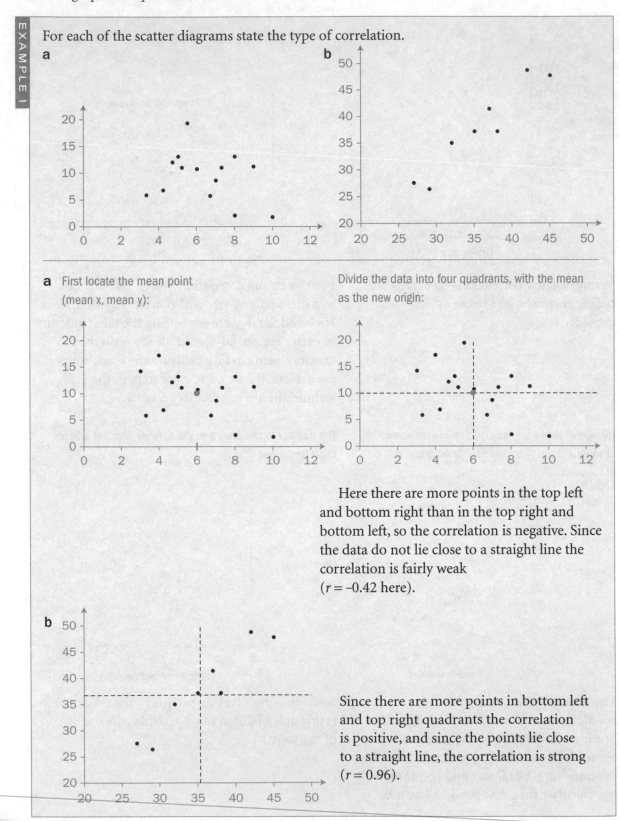

a First locate the mean point
(mean x, mean y):

Divide the data into four quadrants, with the mean
as the new origin:

Here there are more points in the top left
and bottom right than in the top right and
bottom left, so the correlation is negative. Since
the data do not lie close to a straight line the
correlation is fairly weak
($r = -0.42$ here).

b

Since there are more points in bottom left
and top right quadrants the correlation
is positive, and since the points lie close
to a straight line, the correlation is strong
($r = 0.96$).

An alternative way of looking at these scatter diagrams is to sketch an ellipse around the points.

The orientation of the ellipse will tell you whether there is positive or negative correlation, and the relative length and width of the ellipse will tell you something about the strength.

Looking at the previous example:

a

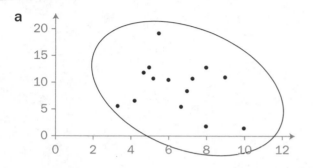

Since the ellipse slopes down from left to right the correlation is negative; since it is not very long and thin the correlation is fairly weak.

b

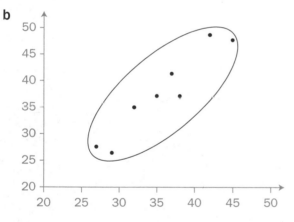

Since the ellipse slopes up from left to right the correlation is positive; since it is long and thin the correlation is strong.

Exercise 5.1

1 **a** For each of the scatter diagrams **i**, **ii** and **iii** state whether or not the product moment correlation coefficient is an appropriate measure to use.

   b State, giving a reason, whether or not the value underneath each diagram is a possible value of the correlation coefficient.

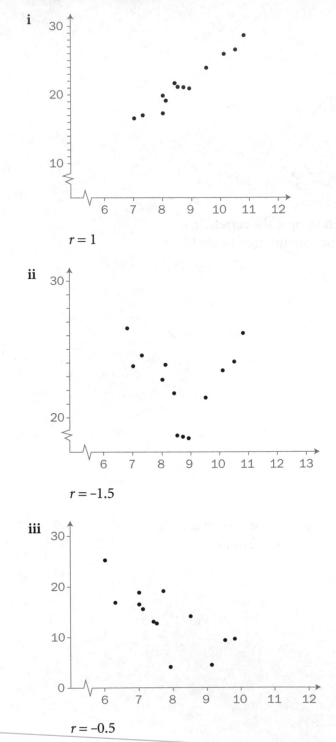

2 a For each of the scatter diagrams **i, ii** and **iii** state whether or
not the product moment correlation coefficient is an appropriate
measure to use.

  b State, giving a reason, whether or not the value underneath
each diagram is a possible value of the correlation coefficient.

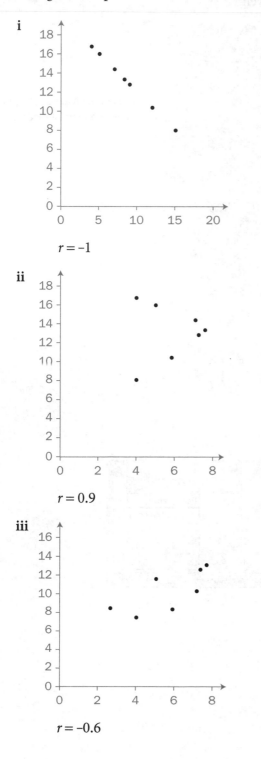

i

$r = -1$

ii

$r = 0.9$

iii

$r = -0.6$

If you know the bivariate data values you can work out the
product moment correlation coefficient (PMCC).
In the Edexcel formulae book, the following are given:

$$S_{xx} = \sum (x_i - \bar{x})^2 = \sum x_i^2 - \frac{(\sum x_i)^2}{n}$$

$$S_{yy} = \sum (y_i - \bar{y})^2 = \sum y_i^2 - \frac{(\sum y_i)^2}{n}$$

$$S_{xy} = \sum (x_i - \bar{x})(y_i - \bar{y}) = \sum x_i y_i - \frac{(\sum x_i)(\sum y_i)}{n}$$

These values are often referred
to as **summary statistics**.

The product moment correlation coefficient is:

$$r = \frac{S_{xy}}{\sqrt{S_{xx} \times S_{yy}}} = \frac{\sum (x_i - \bar{x})(y_i - \bar{y})}{\sqrt{\left\{ \sum (x_i - \bar{x})^2 \right\} \left\{ \sum (y_i - \bar{y})^2 \right\}}}$$

$$= \frac{\sum x_i y_i - \frac{(\sum x_i)(\sum y_i)}{n}}{\sqrt{\left(\sum x_i^2 - \frac{(\sum x_i)^2}{n} \right) \left(\sum y_i^2 - \frac{(\sum y_i)^2}{n} \right)}}$$

Use of a calculator

Almost all standard scientific calculators
now have bivariate data entry. Take
time to learn how to use your
calculator properly.

Often you are given both the original
data and at least some of the summary
statistics. In these cases you can show
the calculation of PMCC using the
formulae, and then check your answer
using the statistical functions on your
calculator.

1.1			RAD AUTO REAL	
A x	B y	C	D	E
			=TwoVar('x,'y,1)	
1	4	5	Statistic	Two-Variable S...
2	7	9	x̄	10.8571
3	2	8	Σx	76.
4	24	9	Σx²	1184.
5	17	34	sx := Sn-...	7.73366

D1 |="Two-Variable Statistics"

S1

EXAMPLE 1

The table shows a set of bivariate data.

x	15	17	20	22	25	28	30	31
y	28.8	26.0	18.5	28.0	24.5	29.5	48.2	41.8

$\sum x = 188$ $\qquad \sum x^2 = 4688$ $\quad \sum y = 245.3$

$\sum y^2 = 8172.67$ $\quad \sum xy = 6040.3$

a Draw a scattergraph to show these data.

b Calculate the correlation coefficient between x and y.

a

b The scattergraph indicates moderate positive correlation – so r should be somewhere in the 0.5–0.75 range.

You could work r out directly using the alternative formula, but then it is more prone to rounding errors, particularly if $\bar{x}$ and $\bar{y}$ are not whole numbers.

$$r = \frac{S_{xy}}{\sqrt{S_{xx} \times S_{yy}}} = \frac{\sum x_i y_i - \frac{(\sum x_i)(\sum y_i)}{n}}{\sqrt{\left(\sum x_i^2 - \frac{(\sum x_i)^2}{n}\right)\left(\sum y_i^2 - \frac{(\sum y_i)^2}{n}\right)}}$$

$$S_{xy} = \sum x_i y_i - \frac{(\sum x_i)(\sum y_i)}{n} = 6040.3 - \frac{188 \times 245.3}{8} = 275.75$$

$$S_{xx} = \sum x_i^2 - \frac{(\sum x_i)^2}{n} = 4668 - \frac{188^2}{8} = 250$$

$$S_{yy} = \sum y_i^2 - \frac{(\sum y_i)^2}{n} = 8172.67 - \frac{245.3^2}{8} = 651.15875$$

$$r = \frac{275.75}{\sqrt{250 \times 651.15875}} = 0.68344\ldots = 0.683 \text{ (3 s.f.)}$$

In the next example, there are ten data pairs so $\bar{x}$ and $\bar{y}$ are straightforward numbers.

Here it is worth using the alternative formula.

You may be given only values of $S_{xx} = \sum(x_i - \bar{x})^2$, $S_{yy} = \sum(y_i - \bar{y})^2$ and $S_{xy} = \sum(x_i - \bar{x})(y_i - \bar{y})$, so you need to be able to work with both formulae.

S1

EXAMPLE 2

S1

Here is a set of bivariate data.

x	6	9	10	12	12	14	17	18	18	20
y	42	47	46	50	44	53	58	51	59	57

Calculate the product moment correlation coefficient and comment on its value.

Put the values in a table:

x	y	$(x_i - \bar{x})$	$(y_i - \bar{y})$	$(x_i - \bar{x})^2$	$(y_i - \bar{y})^2$	$(x_i - \bar{x})(y_i - \bar{y})$
6	42	−7.6	−8.7	57.76	75.69	66.12
9	47	−4.6	−3.7	21.16	13.69	17.02
10	46	−3.6	−4.7	12.96	22.09	16.92
12	50	−1.6	−0.7	2.56	0.49	1.12
12	44	−1.6	−6.7	2.56	44.89	10.72
14	53	0.4	2.3	0.16	5.29	0.92
17	58	3.4	7.3	11.56	53.29	24.82
18	51	4.4	0.3	19.36	0.09	1.32
18	59	4.4	8.3	19.36	68.89	36.52
20	57	6.4	6.3	40.96	39.69	40.32
$\Sigma x = 136$	$\Sigma y = 507$			$S_{xx} = 188.4$	$S_{yy} = 324.1$	$S_{xy} = 215.8$

$$r = \frac{S_{xy}}{\sqrt{S_{xx} \times S_{yy}}} = \frac{215.8}{\sqrt{188.4 \times 324.1}} = 0.873316\ldots = 0.873 \, (3\,\text{s.f.})$$

The variables x and y have strong positive correlation.

A note of caution: Anscombe's data

Some data constructed by Anscombe (1973) illustrate how correlation should be interpreted with care.

The three datasets shown **all** have the same summary statistics:

mean of x-values = 9.0
mean of y-values = 7.5
correlation coefficient = 0.82

When the correlation is high, it is easy to assume that the data look something like this.	These data have the same correlation coefficient, fitting a non-linear relationship perfectly, and a straight line fairly well. Here you should not carry out correlation calculations.	Here, the single outlier has had as much effect as all the small variations in the first case, heavily skewing r.

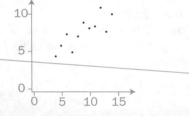
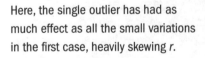

Exercise 5.2

1

x	2.1	2.7	3.5	3.7	4.1	4.8	4.9	5.3	5.9	6.5
y	14.4	12.2	10.6	19.6	19.9	31.0	20.9	22.9	26.0	40.5

$\sum x = 43.5$ $\sum x^2 = 206.65$ $\sum y = 218.0$ $\sum y^2 = 5487.20$ $\sum xy = 1043.62$

a Draw a scatter graph to show these data.

b Calculate the correlation coefficient between x and y.

2

x	46	41	38	36	31	28	25
y	25.1	27.3	29.2	32.0	33.5	37.0	36.5

$\sum x = 245$ $\sum x^2 = 8907$ $\sum y = 220.6$ $\sum y^2 = 7075.44$ $\sum xy = 7522.5$

a Draw a scatter graph to show these data.

b Calculate the correlation coefficient between x and y.

3

x	536	623	654	667	692	739	812	854	892	914
y	-3.5	-3.1	4.1	0.2	-2.9	2.1	-1.7	-0.1	0.8	5.2

$\sum x = 7383$ $\sum x^2 = 5\,592\,735$ $\sum y = 1.1$ $\sum y^2 = 82.11$ $\sum xy = 2553.2$

a Draw a scatter graph to show these data.

b Calculate the correlation coefficient between x and y.

4

x	3.5	4.8	4.9	7.2	8.5	8.8	9.3	10.4	10.8	11
y	0.9	10.9	2.3	0.6	10.9	7.4	7.1	10.4	6.5	2.4

$\sum x = 79.2$ $\sum x^2 = 693.12$ $\sum y = 59.4$ $\sum y^2 = 505.42$ $\sum xy = 499.62$

a Draw a scatter graph to show these data.

b Calculate the correlation coefficient between x and y and comment on its value.

Pupils who are taught in large classes in school often get better results, but this may be because schools tend to put more able pupils in large groups so that teachers can offer greater support to weaker pupils in smaller classes.

Correlation does not imply causation. Both variables may really be tied to a third (hidden or confounding) variable, particularly size or time, for example.

Here are some other considerations that you should be aware of.

- **Controlling the conditions more closely will strengthen the correlation**
  In measuring the effects of fertiliser, monitoring all the fields on a single farm will give a stronger correlation than a sample of fields in farms across the country.

  However, this will mean that knowledge gained from the data will only be useful in similarly controlled conditions.

- **Different samples from the same population are likely to give different values of r**
  This variation decreases as the sample size increases. Therefore it is not sensible to report the value of r to lots of decimal places.

- **Outliers can make a huge difference to r**
  This is part of the reason why drawing a scatter graph is always a good idea.

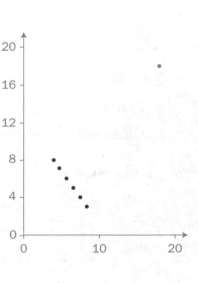

In the scatter graph on the right without the blue point there is perfect negative correlation i.e. $r = -1$.
With it $r = 0.75$.

EXAMPLE 1

The marks in Maths, Physics and Art are collected from a sample of students.

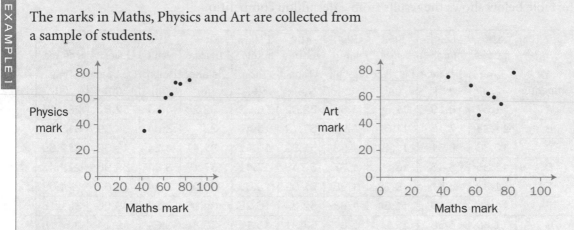

Comment on the following claims:

a Pupils who do well in Maths tend to do well in Physics.

b Pupils who do well in Maths tend to do well in Art.

---

a The marks in Maths and Physics show strong positive correlation so, for these students at least, those who do well in Maths tend to do well in Physics.

b The marks in Maths and Art show little correlation so, for these students at least, there is nothing to suggest that those who do well in Maths will do well in Art.

Exercise 5.3

1 The average temperature and rainfall were collected for a number of cities around the world.

Temp (°C)	17	18.7	24.4	22.8	20.7	11.8	18.5	24.7	26	34.6	16.7	27.9	28.5	16.3
Rainfall (mm)	59	236.8	16.3	83.6	72.3	102.8	81.5	135.6	224.3	0	0.9	0	158.2	4.4

The scatter graph shows this information.

a Comment on the claim that hotter cities have less rainfall.

b Calculate the correlation coefficient between average temperature and average rainfall.

(You could practice using the bivariate data functionality on your calculator).

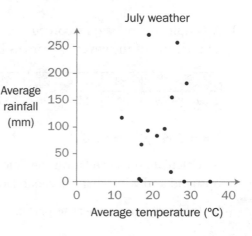

2   The table below shows the results from a decathlon competition.

Competitor	100 metres Time (s)	Long jump Distance (m)	Shot putt Distance (m)	High jump Height (m)	400 metres Time (s)	110 m hurdles Time (s)	Discus Distance (m)	Pole vault Height (m)	Javelin Distance (m)	1500 metres Time (s)
A	11.57	6.19	11.56	1.89	52.62	15.52	32.24	4.54	52.82	280.14
B	11.28	7.15	11.6	1.95	52.19	15.49	33	3.64	54.09	313.27
C	10.86	6.61	11.42	1.83	51.38	15.41	31.47	4.24	48.02	312.69
D	11.35	5.46	13.2	1.77	52.48	17.41	37.15	3.64	53.96	272.47
E	11.33	6.51	9.78	1.95	51.09	16.65	28.51	3.14	52.69	274.09
F	11.46	6.69	11.64	1.83	52.28	16.28	34.96	3.84	49.06	321.18
G	11.43	6.56	12.1	1.8	54.31	15.1	30.85	3.64	49.14	367.18
H	11.38	6.46	12.01	1.62	55.68	15.67	33.35	3.84	41.79	309.21
I	11.95	6.57	11.78	1.95	57.07	17.47	30.43	3.44	42.11	355.24
J	11.45	5.89	9.89	1.8	55.66	16.72	35.93	2.84	40.33	346.71

Good performances in the 100 metres are the low times and good performances in the javelin are the long distances, so if athletes tend to be good in both disciplines the correlation would be negative.

a   i   Calculate the correlation coefficient between the performances in the 100 metres and the javelin.
    ii Interpret your result.

Try using your calculator to give you the summary statistics.

b   i   Calculate the correlation coefficient between the performances in the discus and the long jump.
    ii Interpret your result.

You might like to think in which pairs of events you would expect to see positive, negative or little correlation and investigate in a spreadsheet or statistical software package.

3   A sample of marriages recorded at a registry gave the ages of the husbands, h years, and of the wives, w years, as in the table below.

h	22	23	27	27	28	31	33	34	41
w	23	21	25	26	31	31	32	30	29

$\sum h = 266$ $\sum h^2 = 8142$ $\sum w = 248$ $\sum w^2 = 6958$ $\sum hw = 7460$

a   Calculate the correlation coefficient between the ages of husband and wife when they married.

b   Interpret your answer to part a.

4 The eruptions of the Old Faithful geyser are monitored regularly. Table 1 shows the duration of an eruption, d minutes, and the interval, i minutes, until the next eruption.

Table 1

Duration d	4.3	2.6	4.3	3.7	2.9	4.8	5	1.9	4.6	1.7	3.5
Interval i	75	75	74	69	73	43	52	80	69	76	78

$\sum d = 39.3 \quad \sum d^2 = 153.79 \quad \sum i = 764 \quad \sum i^2 = 54390 \quad \sum di = 2640.7$

a i Calculate the correlation coefficient between d and i.

ii Interpret your answer to part **ai**.

Data is also collected on the intervals between eruptions to help park staff predict when the next eruption might occur. Table 2 shows a sample of times of an interval, f minutes, between eruptions and the time, n minutes, of the following interval between eruptions.

Table 2

First interval f	51	90	77	51	86	74	52	44
Next interval n	82	57	77	95	53	77	81	89

$\sum f = 525 \quad \sum f^2 = 36743 \quad \sum n = 611 \quad \sum n^2 = 48147 \quad \sum fn = 38470$

b i Calculate the correlation coefficient between f and n.

ii Interpret your answer to part **bi**.

5 The CensusAtSchool project has collected data from school pupils across the UK about various topics. A sample of pupils was taken from the database. The table below shows the ages, x years, and the times of going to bed the previous night, y, given as a decimal time in the 24-hour clock, i.e. 21.30 means 9.30 pm.

x	7	8	9	10	11	12	13	14	15	16
y	19.75	20.00	22.50	21.67	23.50	23.00	22.00	23.00	23.00	23.00

$\sum x = 115 \quad \sum x^2 = 1405 \quad \sum y = 221.42 \quad \sum y^2 = 4918.15 \quad \sum xy = 2572.95$

a Draw a scatter graph to represent this information.

b Calculate the correlation coefficient between x and y.

c Caleb claims older children go to bed later. Comment on this claim, making reference to your answer to part **b**.

A **linear** change in scale of x or y will make no difference to the correlation coefficient.

Here is a set of bivariate data, with $r = 0.93$.

Here, all the y values have had 5 subtracted. Note that r is unchanged.

Each y value has been doubled and then 10 has been subtracted. Note that the gradient is steeper but r is still unchanged.

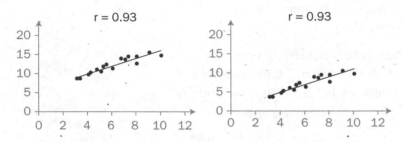

You can use this fact to make the calculation of r easier, by transforming the x and y values to easier numbers.

So in question 1 below, your answers to parts **b** and **c** should be the same.

Exercise 5.4

1 A golfer wants to buy some new golf balls and decides to try seven different types to see how far they travel with his clubs. He records the distance l metres he hits each type of ball with a 4-iron, and the distance m metres he hits it with a 9-iron.

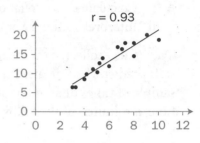

	l	*m*
A	167	119
B	163	116
C	157	113
D	169	122
E	155	114
F	163	117
G	160	116

When these data are coded using $x = \dfrac{l - 150}{3}$ and $y = \dfrac{m - 110}{2}$,

$\Sigma x = 28$ $\Sigma y = 23.5$ $\Sigma x^2 = 129.11$

$\Sigma y^2 = 92.75$ $\Sigma xy = 108.5$

a Calculate S_{xy}, S_{xx} and S_{yy}.

b Calculate, to three significant figures, the value of the product moment correlation coefficient between x and y.

c Calculate the value of the product moment correlation coefficient between l and m.

2 The average temperature and rainfall were collected for a number of cities round the world.

Temp (°C)	17	18.7	24.4	22.8	20.7	11.8	18.5	24.7	26	34.6	16.7	27.9	28.5	16.3
Rainfall (mm)	59	236.8	16.3	83.6	72.3	102.8	81.5	135.6	224.3	0	0.9	0	158.2	4.4

a Change the units of temperature from Celsius to Fahrenheit (multiply by 1.8 and add 32), and of rainfall from mm to inches (divide by 2.54).

b Calculate the correlation coefficient of the data in Fahrenheit and inches.

c State the correlation coefficient of the data in Celsius and mm.

3 Dunnocks (or hedge sparrows) are one species of small birds found in Great Britain. The table below shows the wingspan, x cm, and the weight, y grams, of a sample of dunnocks.

Wingspan, x (cm)	67	68	71	72	69	71	71	71	68	72
Weight, y (g)	19.7	23.9	22.8	23.4	20.8	23.7	19.5	21	21.8	22.8

a Code these data by $X = x - 60$ and $Y = y - 18$.

b Calculate the correlation coefficient between X and Y.

$$\Sigma X = 100 \quad \Sigma X^2 = 1030 \quad \Sigma Y = 39.4 \quad \Sigma Y^2 = 178.96 \quad \Sigma XY = 402.1$$

c State the value of the correlation coefficient between the wingspan and the weight of dunnocks.

4 Blue tits are small birds. Chris claims they gain weight during the day so that they have some fat to burn off during the cold night. The table below shows the weights, w grams, of a sample of birds and the time, t, recorded as a decimal in hours, at which the measurement was taken.

Time, t (h)	9.9	10.2	11.3	11.1	11.8	10.8	11.4	9.3	10.9	10.3
Weight, w (g)	10	11	11	10	11	8	12	9	6	13

a Code these data by $T = t - 9$ and $W = w - 5$.

b Calculate the correlation coefficient between T and W.

$$\Sigma T = 17 \quad \Sigma T^2 = 34.18 \quad \Sigma W = 51 \quad \Sigma W^2 = 297 \quad \Sigma TW = 88.5$$

c Comment on Chris' claim.

d How could the data collection have been improved to make it easier to investigate Chris' claim?

1 The scatter diagrams below were drawn by a student.

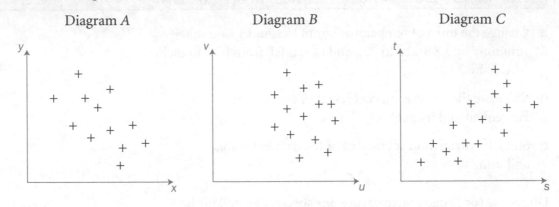

Diagram A   Diagram B   Diagram C

The student calculated the value of the product moment correlation coefficient for each of the sets of data.

The values were   0.68   −0.79   0.08

Write down, with a reason, which value corresponds to which scatter diagram.

[Edexcel, 2005]

2 A company owns two petrol stations P and Q along a main road. Total daily sales in the same week for P ($£p$) and for Q ($£q$) are shown in the table.

When these data are coded using

$$x = \frac{p - 4365}{100} \text{ and } y = \frac{q - 4340}{100},$$

$\Sigma x = 48.1$ $\Sigma y = 52.8$ $\Sigma x^2 = 486.44$
$\Sigma y^2 = 613.22$ and $\Sigma xy = 204.95$.

	p	q
Monday	4760	5380
Tuesday	5395	4460
Wednesday	5840	4640
Thursday	4650	5450
Friday	5365	4340
Saturday	4990	5550
Sunday	4365	5840

a Calculate S_{xy}, S_{xx} and S_{yy}.

b Calculate, to three significant figures, the value of the product moment correlation coefficient between x and y.

c i Write down the value of the product moment correlation coefficient between p and q.

   ii Give an interpretation of this value.

3 In order to assess the wearing characteristics of tyres, six motorcycles were fitted with new tyres, ridden for 1000 miles, and the depth of tread on each tyre was measured. The results are shown below.

Motorcycle	A	B	C	D	E	F
Front tread, x	2.4	1.9	2.3	1.9	2.4	2.5
Back tread, y	2.0	2.1	2.0	1.9	1.8	2.0

The data in the table can be summarised as follows.

$$\Sigma x = 13.4 \quad \Sigma y = 11.8 \quad \Sigma x^2 = 30.28 \quad \Sigma y^2 = 23.26 \quad \Sigma xy = 26.32$$

a Calculate the correlation coefficient between x and y.

b Comment briefly on its value.

Stefan thinks that they should have been using the tread which had been worn away in the 1000 miles, so he codes the measurements using $X = 3 - x$, $Y = 3 - y$.

c Find the correlation coefficient between X and Y.

4 Hedge funds aim to provide good financial returns for investors which are uncorrelated with the performance of equity markets, allowing the investors to diversify their risk. The BennEx Hedge Fund is based in Hong Kong.
The annualised monthly returns for 11 months for the BennEx Hedge Fund, x %, and for the main Hong Kong equity market, y %, are shown in the table below:

x	11.4	8.2	14.5	9.6	8.3	11.3	7.3	9.8	12.4	6.1	10.7
y	7.8	8.7	9.3	8.2	9.1	7.8	6.3	5.7	7.3	7.5	8.6

a On graph paper, draw a scatter diagram to represent these data.

The data in the table can be summarised as follows.

$$\Sigma x = 109.6 \quad \Sigma y = 86.3 \quad \Sigma x^2 = 1150.98 \quad \Sigma y^2 = 689.59 \quad \Sigma xy = 867.64$$

b Calculate the correlation coefficient between x and y.

c Annie says that the returns should be the actual monthly returns rather than the annualised returns, and wants to code the data using $X = \frac{x}{12}$, $Y = \frac{y}{12}$.
Find the correlation coefficient of X and Y.

d Comment on whether the BennEx Hedge Fund performance satisfies the aim of not being correlated with the Hong Kong market.

S1

5 The table below shows the amount of fertiliser used on plots of land, x, and the crop yield, y.

x	20	35	42	46	55	63	69	74	78
y	27	41	41	37	41	47	55	59	58

The data in the table can be summarised as follows.

$$\Sigma x = 482 \quad \Sigma y = 406 \quad \Sigma x^2 = 28\,820 \quad \Sigma y^2 = 19\,220 \quad \Sigma xy = 23\,300$$

a Calculate the correlation coefficient.

b Comment briefly on its value.

c Catharine says that the data do not make sense because there are no units provided. Explain why knowing the units will make no difference to the value of the correlation coefficient.

6 A physiotherapist is interested to know what she should expect normal people to do on various tasks she uses in assessing muscle injuries. She decided to time a number of healthy students, each carrying out the same task once with the left hand, and once with the right hand.
The times, in seconds, were as follows:

Student	P	Q	R	S	T	U	V	W
Left hand, x	35	38	39	43	44	44	47	48
Right hand, y	26	29	26	31	29	34	33	36

The data in the table can be summarised as follows.

$$\Sigma x = 338 \quad \Sigma y = 244 \quad \Sigma x^2 = 144\,240 \quad \Sigma y^2 = 7536 \quad \Sigma xy = 10\,410$$

a Calculate the correlation coefficient.

b Comment briefly on its value.

7 The CensusAtSchool project has data on whether the pupils trust a variety of institutions such as the police and the press.
The table below shows the level of trust (on a continuous scale of 0 to 5) of the police, x, and the press, y, for a sample of pupils.

x	2.07	3.19	3.41	1.52	3.98	1.03	0.13	3.89	3.39	0.83
y	1.07	1.96	2.93	2.66	0.77	0	0.35	1.14	0.35	0.13

$$\Sigma x = 23.44 \quad \Sigma x^2 = 72.63 \quad \Sigma y = 11.36 \quad \Sigma y^2 = 22.80 \quad \Sigma xy = 31.34$$

a Draw a scatter graph to represent this information.

b Calculate the correlation coefficient between x and y.

c Interpret your answer to part b.

8 In a television game show, the contestant has to guess the value of various prizes.
The actual value, v, and the guess, g, for each prize are given in the table.

(You may assume that $\Sigma v = 14\,300$ $\Sigma g = 22\,480$
$\Sigma vg = 240\,916\,400$ $\Sigma v^2 = 144\,911\,800$
$\Sigma g^2 = 400\,992\,400$)

Prize	v	g
MP3 player	120	90
Washing machine	350	420
Television	400	500
Computer	600	450
Weekend break for two	400	350
Garden table + four chairs	250	400
Watch	180	270
Car	12000	20000

a Draw a scatter diagram.

b Calculate, to four significant figures, the product moment correlation coefficient for the above data.

c Explain why the value of the product moment correlation coefficient would be lower if the car was excluded from the data.

5

Exit →

Summary

Refer to

- The product moment correlation coefficient, r, is a measure of how well the data fit a straight line.

  5.1

- r can be calculated using the statistical functions on a calculator, or by using summary statistics.

  - $$S_{xx} = \sum(x_i - \overline{x})^2 = \sum x_i^2 - \frac{(\sum x_i)^2}{n}; \quad S_{yy} = \sum(y_i - \overline{y})^2 = \sum y_i^2 - \frac{(\sum y_i)^2}{n}$$

    $$S_{xy} = \sum(x_i - \overline{x})(y_i - \overline{y}) = \sum x_i y_i - \frac{(\sum x_i)(\sum y_i)}{n}$$

  - $$r = \frac{S_{xy}}{\sqrt{S_{xx} \times S_{yy}}}$$

  5.2

- $|r| >$ about 0.75 is generally described as strong correlation, $|r| <$ about 0.4 is generally described as weak and in between these values as moderate correlation – and any description should also refer to whether the correlation is positive or negative. You should describe relationships in context where possible – positive correlation means that high values of the two variables tend to go together as do low values of both variables, and negative correlation occurs if one variable is high when the other is low.

  5.3

- Any linear coding of the x- and/or the y-values leaves the value of the correlation coefficient unchanged.

  7.4

Links

In the financial services industry, one of the primary objectives in building a portfolio of investments is to diversify risk so that a downturn in one part of the portfolio does not mean that other parts will also do badly – and so fund managers actively seek *market neutral* investments which are not correlated with traditional components of investment portfolios such as equities, property, gilts etc.

In medicine, data logging means that a lot of extra information can be collected from patients. Data mining techniques then trawl these data looking for associations which medical staff can then investigate to see if there are causal links.

1 An outline of the stages in developing a statistical model are
listed below with stages 2 and 5 missing.

Stage 1   The recognition of a real-life problem.
Stage 2
Stage 3   The model is used to make predictions
Stage 4   Experimental or observational data is collected
Stage 5
Stage 6   Statistical techniques are used to test how well the
           model describes the real-life problem.
Stage 7   Refine the model.

Suggest what stages 2 and 5 are.   (2)

2 The total amount of time the senior editor of a publishing firm spent
in meetings in a working day was recorded to the nearest minute.
The data collected over 40 days are summarised in the table below.

Time (min)	90–139	140–149	150–159	160–169	170–189
No. of days	8	10	14	4	4

a Draw a histogram to illustrate these data (use the same
   scales as in the histogram shown below).   (4)

b Calculate estimates of the mean and standard deviation
   of the time per day the senior editor spends in meetings.   (4)

The publishers are producing a major new suite of A-level texts.
The histogram below illustrates the time the senior editor spent
in meetings during the 40 working days before the deadline
for handing over the suite of texts.

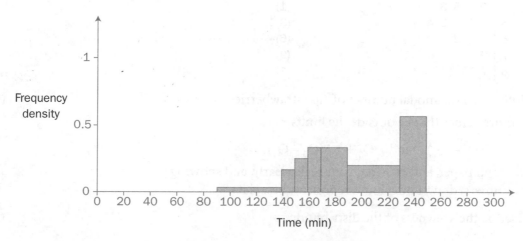

c Comment on the differences in the times the senior editor
   spent in meetings in the two periods.   (3)

3 The heights, h, of a group of ten plants from one garden has

$$\Sigma h = 93.7 \text{ cm} \quad \Sigma h^2 = 1071.26 \text{ cm}^2$$

  a Calculate the mean and standard deviation of the heights
     of these plants. (4)

A group of 40 plants of the same species from another garden
has mean height 9.81 cm and standard deviation 5.2 cm.

  b Compare the distributions of the heights of the plants
     from the two gardens. (2)

4 The times taken by people at breakfast in a hotel are
  summarised in the table.

Time, t minutes, taken for breakfast	Number of people
$0 \leqslant t < 5$	12
$5 \leqslant t < 10$	37
$10 \leqslant t < 15$	34
$15 \leqslant t < 20$	9

Mr Fisher, the hotel manager, uses $x = \dfrac{t - 2.5}{5}$.

  a Show that $\Sigma x = 132$ and find Σx^2. (4)

  b Calculate estimates of the mean and standard deviation
     of the times taken for breakfast in the hotel. (6)

5 The stem and leaf diagram shows the number of ripe strawberries
  picked in 22 rows of strawberry plants.

Number of ripe strawberries   3 | 9 means 39 strawberries

```
3 | 8 9                          (2)
4 | 0 1 3 3 5 5 5 7 9            (1)
5 | 0 0 1 3 4 5                  (3)
6 | 1 7 8                        (6)
7 | 9                            (9)
8 | 2                            (1)
```

  a Write down the modal number of ripe strawberries in a row. (1)

Outliers are values that lie outside the limits

$$Q_1 - 1.5(Q_3 - Q_1) \quad \text{and} \quad Q_3 + 1.5(Q_3 - Q_1).$$

  b On graph paper, and showing your scale clearly, and showing
     any outliers, draw a boxplot to represent these data. (8)

  c Describe the skewness of the distribution. (1)

6 Summarised below are the data relating to the number of minutes, to the nearest minute, that a random sample of 65 trains from Darlingborough arrived late at a main line station.

Minutes late (0 | 2 means 2 min)		Totals
0	2 3 3 3 4 4 4 4 5 5 5 5 5 5	(14)
0	6 6 6 7 7 8 8 8 9	(9)
1	0 0 0 2 2 3 3 4 4 4 5	(a)
1	6 6 7 7 8 8 8 9 9	(b)
2	1 2 2 3 3 3 3 4	(c)
2	6	(d)
3	3 4 4 5	(4)
3	6 8	(2)
4	1 3	(2)
4	7 7 9	(3)
5	2 4	(2)

a Write down the values (a)–(d) needed to complete the stem and leaf diagram. (4)

b Find the median and quartiles of these times. (4)

c Find the 67th percentile. (2)

d On graph paper construct a boxplot for these data, showing your scale clearly. (4)

e Comment on the skewness of the distribution. (1)

A random sample of trains arriving at the same main line station from Shefton had a minimum value of 15 minutes late and a maximum value of 30 minute late. The quartiles were 18, 22 and 27 minutes.

f On the same graph paper and using the same scale, construct a boxplot for these data. (3)

g Compare and contrast the train journeys from Darlingborough and Shefton based on these data. (2)

7 The children in classes A and B were each given a set of arithmetic problems to solve. Their times, to the nearest minute, were recorded and they are summarised in the table below.

	Class A	Class B
Smallest value	5	10
Largest value	27	26
Q1	9	13
Q2	15	15
Q3	18	22

a On graph paper and using the same scale for both, draw boxplots to represent these data. (6)

b Compare and contrast the results for the two classes. [(c) Edexcel Limited 1997] (3)

8 The following table summarises the results of a sales manager's analysis of the amounts to the nearest £, of a sample of 750 invoices.

Amount of invoice (£), x	Number of invoices, f
0–9	50
10–19	204
20–49	165
50–99	139
100–149	75
150–199	62
200–499	46
500–749	9

Let x represent the midpoint of each class.

Thus with $y = \frac{x - 14.5}{10}$, $\Sigma fy = 5021$ and $\Sigma fy^2 = 115\,773.5$

a Using these values, or otherwise, find estimates of the mean and the standard deviation of the population from which this sample was taken. (4)

b Explain why the mean and the standard deviation might not be the best summary statistics to use with these data. (2)

c Calculate estimates of alternative summary statistics which could be used by the sales manager. (5)

d Use these estimates to justify your explanation in part b. [(c) Edexcel Limited 1997] (2)

9 The 30 members of the Darton town orchestra each recorded the amount of individual practice, x hours, they did in the first week of June. The results are summarised as follows:

$$\sum x = 225 \qquad \sum x^2 = 1755$$

The mean and standard deviation of the number of hours of practice undertaken by the members of the Darton orchestra in this week were μ and σ respectively.

a Find μ. (2)

b Find σ. (3)

Two new people joined the orchestra and the number of hours of individual practice they did in the first week of June were $\mu - 2\sigma$ and $\mu + 2\sigma$.

c State, giving your reasons, whether they effect of including these two members increases, decreases, or leaves unchanged the mean and the standard deviation. [(c) Edexcel Limited 1997] (3)

10 The mark, x, obtained by each of 45 students randomly selected from those students who sat the accountancy examination was recorded. The stem and leaf diagram below summarises the marks.

Mark (5 | 3) means 53)		Total
5	0 1 3 3 4 4	(6)
5	5 6 7 9	(4)
6	1 1 3 3 4 4 4	(7)
6	5 7 8 8 9	(5)
7	3 3 4 4 4 4	(6)
7	5 5 6 6 7 7 7 7 8 8 8 9 9	(13)
8	0 0 1 1	(4)

a Using graph paper and showing your scale clearly, construct a boxplot to represent these data. (8)

b Comment on the skewness of this distribution. (1)

For the above sample $\sum x = 3085$ and $\sum x^2 = 215\,569$

c Find the mean and the standard deviation of this sample of marks. (3)

The mean and the standard deviation of the marks of all the students who sat the examination were 65 and 16.5 respectively. The examiners decided that the mark of each student should be scaled by having 10 marks subtracted and then reduced by a further 10%.

d Find the mean and the standard deviation of the scaled marks. [(c) Edexcel Limited 2000] (5)

S1

11 Two identical bags each contain 12 discs, which are identical except for colour. One bag (A) contains six red and six blue discs, and the other (B) contains eight red and four blue discs.

    a A bag is selected at random and a disc is selected from it. Calculate the probability that the disc drawn will be red. (2)

    b The disc selected is now returned to the same bag, along with another two of the same colour, and another disc is chosen from that bag. Find the probability that:

      i it is the same colour as the first disc drawn
      ii bag A was used, given that two discs of the same colour have been chosen. (4)

12 In a game of chess a player can win, lose or draw and the likelihood of these events depends on whether the player uses the white or black pieces. Mr Todd and Mr du Feu play chess regularly against one another.
If Mr Todd is playing with the white pieces he wins with probability 0.6, and draws with probability 0.1. If he is playing with the black pieces he wins with probability 0.4, and draws with probability 0.3.
They toss a coin to see who plays with the white pieces first, but then they alternate colours.

    a Draw a tree diagram to show the possibilities of the first two games played in an evening. (4)

    b What is the probability that Mr Todd wins the first game played? (2)

    c What is the probability that Mr Todd wins both of the first two games played? (3)

    d Given that Mr Todd wins exactly one of the two games played, what is the probability it was the game in which he had the black pieces? (5)

13 A child has a bag containing 12 sweets of which 3 are yellow, 5 are green and 4 are red. When the child wants to eat a sweet, a random selection is made from the bag and the chosen sweet is then eaten before the next selection is made.

    a Find the probability that the child does not select a yellow sweet in the first two selections.

    b Find the probability that there is at least one yellow sweet in the first two selections.

    c Find the probability that the third sweet selected is yellow, given that the first two sweets selected were red ones. [(c) Edexcel Limited specimen] (11)

14 There are 60 students in the sixth form of a certain school. Mathematics is studied by 27 of them, Biology by 20 of them and 22 students study neither Mathematics nor Biology.

   a Find the probability that a randomly selected student studies both Mathematics and Biology.

   b Find the probability that a randomly selected student does not study Biology.

   A student is selected at random.

   c Determine whether the event 'studying Mathematics' is statistically independent of the event 'not studying Biology'.   [(c) Edexcel Limited 1996] (10)

15 The three events E_1, E_2 and E_3 are defined in the same sample space. The events E_1 and E_3 are mutually exclusive, the events E_1 and E_2 are independent.

   Given that $P(E_1) = \frac{2}{5}$, $P(E_3) = \frac{1}{3}$ and $P(E_1 \cup E_2) = \frac{5}{8}$ find

   a $P(E_1 \cup E_3)$ (3)

   b $P(E_2)$ [(c) Edexcel Limited 1998] (3)

16 A market researcher asked 100 adults which of the three newspapers, A, B, and C, they read. The results showed that 30 read A, 26 read B, 21 read C, 5 read both A and B, 7 read both B and C, 6 read both C and A and 2 read all three.

   a Draw a Venn diagram to represent these data.   (6)

   One of the adults is then selected at random.

   Find the probability that she reads

   b at least one of the newspapers   (2)

   c only A   (1)

   d only one of the newspapers   (2)

   e newspaper A, given that she reads only one newspaper.   [(c) Edexcel Limited 2001] (2)

S1

17 A keep fit enthusiast swims, runs or cycles each day with probabilities 0.2, 0.3 and 0.5 respectively. If he swims he then spends time in the sauna with probability 0.35. The probabilities that he spends time in the sauna after running or cycling are 0.2 and 0.45 respectively.

    a Represent this information on a tree diagram. (3)

    b Find the probability that on any particular day he uses the sauna. (3)

    c Given that he uses the sauna one day, find the probability that he has been swimming. (3)

    d Given that he does not use the sauna one day, find the probability that he has been swimming. [(c) Edexcel Limited practice paper] (3)

6

Regression

This chapter will show you how to
- calculate the equation of a line of best fit from raw bivariate data and from summary statistics
- make predictions or estimates of values of the dependent variable from a value of the independent variable
- work with coded bivariate data.

Before you start

You should know how to:

1 Calculate the mean of a set of data.

e.g. Students enrolled on an Italian language course were asked how many times they had visited Italy. The results are shown in the table.

No. of visits	0	1	2	3	4	5	8
Frequency	3	2	5	8	2	1	1

Find the mean number of visits.

Total number of visits $= (0 \times 3) + (1 \times 2) + (2 \times 5)$
$+ (3 \times 8) + (4 \times 2)$
$+ (5 \times 1) + (5 \times 1)$
$= 57$

Total number of students $= 3 + 2 + 5 + 8 + 2 + 1 + 1$
$= 22$

Mean number of visits $= 57 \div 22$
$= 2.60$ (to 2 d.p.)

2 Substitute values into simple equations.

e.g. If $c = a^2 + 2ab$,

find c when $a = 3.1$ and $b = -2.4$.

Substitute for a and b:

$c = 3.1^2 + 2(3.1 \times -2.4)$
$= 9.61 - 14.88$
$= -5.27$

Check in

1 Calculate the mean of 5, 7, 8, 10, 13.

2 Find y when $x = 8.3$ given that
$y = 3.02x - 7.1$

A mail order company manager records the time taken by an employee to complete a number of orders. The results are shown in the graph below.

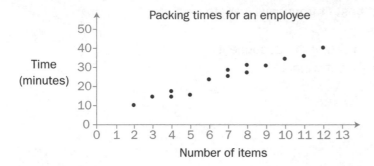

It is clear that the time taken depends on the number of items in the order. It looks as if a straight line graph could be used to give quite good estimates.

> A line of best fit is often known as a regression line.
> In bivariate data, if one of the variables is controlled, it is known as the independent variable. If the other variable is measured, it is known as the dependent variable.

Time is the dependent or response variable and the number of items is the independent or explanatory variable.

In science experiments, you often vary one factor and measure the change to another factor. These are the independent and dependent variable respectively.

You could draw a straight line to fit the data, like the one below. The independent variable is plotted an the horizontal axis.

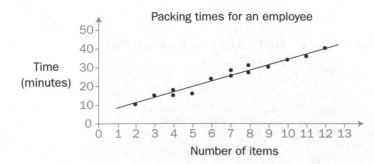

There are other factors, such as the size of the items, which mean that packing times are not perfectly predictable, but $y = ax + b$ is a fairly accurate model.

From the straight line, you could predict the time taken for a given number of items.

However, if the data are not so strongly correlated, it will not be as easy to decide which line is the 'best fit' to the data.

The graph below shows the packing times of a sample of orders from all employees.

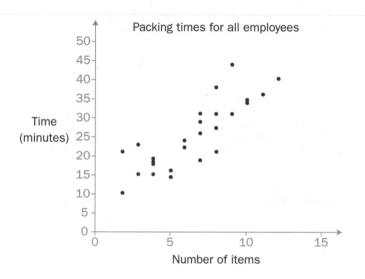

Now it is not as easy to predict the packing time from the number of items. If you draw an ellipse around the observations and take the longer axis, it will give you an approximate line of best fit.

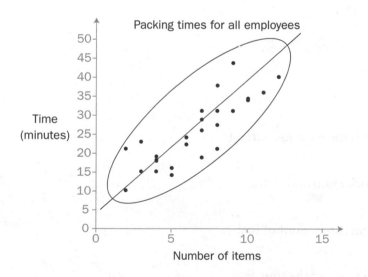

If you use this line for prediction, it will give you less reliable results than if the points were closer to a straight line.

Models of predictability do not need to be linear.

The following graph shows the temperature of some cakes at various times after they have been taken out of the oven.

The temperature seems to be reasonably predictable, but not by fitting a straight line to the observations.

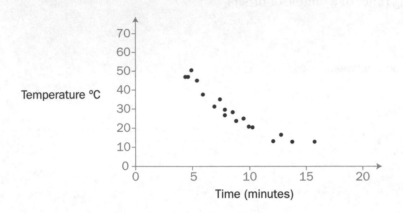

It would make more sense to fit a decreasing curve to the data points.

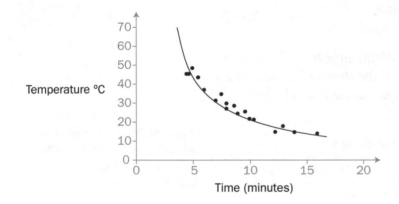

Exercise 6.1

1 For each of the following, state which is the response variable and which is the explanatory variable:

   a the marks a student scores in a mock exam and in the final exam

   b the amount of energy produced and the amount of fuel used in a boiler

   c the distance an athlete runs in training and the time taken to do it

   d the cost of a gold necklace and its weight.

SI

2 For each of the following scatter graphs state whether it is
reasonable to fit a straight line to the data. Give a reason for any
graph for which you do not think a line of best fit is suitable.

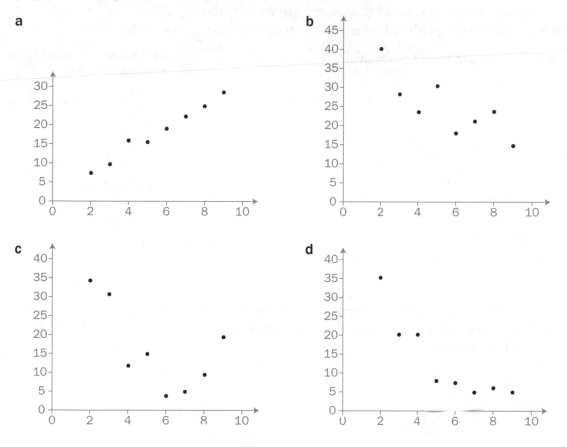

a

b

c

d

When you fit a straight line model to data, you can measure the difference between the 'predicted value' and the observed value.

These differences are known as **residuals**.

The residuals r_i are shown by the vertical broken lines.
$$r_i = y_i - a - bx_i$$
where $y = bx + a$ is the equation of the line of best fit.

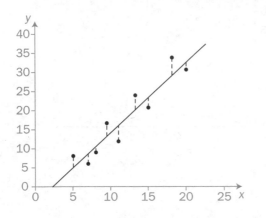

If you construct the straight line so that the sum of the squares of the residuals is as small as it could possibly be, this will give you a very good line of best fit.

If you just added the residuals, the positives and negatives would cancel each other out. So you add the squares of the residuals instead.

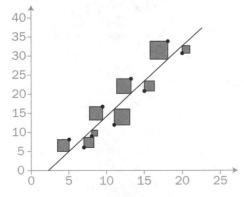

The **least squares line of regression** for a set of data is found by calculating the minimum sum of the squares of the

To find the equation of the least squares line of regression, you need to use:

- algebra – you know that the equation of a straight line is
  $$y - y_1 = m(x - x_1)$$
- calculus – you can use differentiation to find minimum values.

S1

The derivation is slightly beyond the scope of the course, but here are the results.

> The least squares line of regression is given by
> $y - \bar{y} = b(x - \bar{x})$ where
> $$b = \frac{\sum(x_i - \bar{x})(y_i - \bar{y})}{\sum(x_i - \bar{x})^2} = \frac{\sum x_i y_i - \frac{(\sum x_i)(\sum y_i)}{n}}{\sum x_i^2 - \frac{(\sum x_i)^2}{n}} = \frac{S_{xy}}{S_{xx}}$$

If $y = a + bx$ is the least squares line, then $a = \bar{y} - b\bar{x}$.

You can find the equation of a regression line:
- directly from the data, using your calculator
- from the summary statistics ($\sum x_i y_i$ etc).

Once you have the equation of the regression line, you can:
- plot it onto the scatter graph as a line of best fit
- use the equation directly to calculate predicted values.

It is worth remembering that $(\bar{x}, \bar{y})$ always lies on any line of regression.

EXAMPLE 1

The variables C and R are known to be approximately linearly related. Seven pairs of values (c, r) of the variables gave the following results:

$\sum c = 58$, $\sum r = 51$, $\sum c^2 = 530$, $\sum r^2 = 379.76$, $\sum cr = 402.7$

a Find the values of $\bar{c}$, $\bar{r}$, S_{cr}, S_{cc}

b Find the equation of the regression line of r on c.

---

a $\bar{c} = \dfrac{58}{7}$

$\bar{r} = \dfrac{51}{7}$

$S_{cr} = \sum cr - \dfrac{(\sum c)(\sum r)}{n} = 402.7 - \dfrac{58 \times 51}{7} = -19.8714\ldots$

$S_{cc} = \sum c^2 - \dfrac{(\sum c)^2}{n} = 530 - \dfrac{58^2}{7} = 49.4285\ldots$

b $b = -\dfrac{19.8714\ldots}{49.4285\ldots} = -0.402023\ldots$

Remember:
$b = \dfrac{S_{xy}}{S_{xx}}$
In this case $b = \dfrac{S_{cr}}{S_{cc}}$

$r - \bar{r} = -0.402(c - \bar{c})$ $\therefore r = -0.402c + \left(0.402 \times \dfrac{58}{7}\right) + \dfrac{51}{7}$

$= -0.402c + 10.6$

So the equation of the regression line is
$r = -0.402c + 10.6$

Exercise 6.2

1 For the following trivial data sets, use the data to check that the summary statistics are correct.

a

x	1	2	3
y	5	7	9

$\sum x = 6$ $\sum x^2 = 14$ $\sum y = 21$ $\sum y^2 = 155$ $\sum xy = 46$

$a = 3$ $b = 2$

b

x	1	2	3
y	5	6	9

$\sum x = 6$ $\sum x^2 = 14$ $\sum y = 20$ $\sum y^2 = 142$ $\sum xy = 44$

$a = 2.667$ $b = 2$

2 Plot the data below on a scatter graph.

x	14	16	21	25	32	37	46	54	60	72	78	81
y	34	36	42	46	51	57	68	74	81	90	92	97

Find the equation of the regression line of y on x.
You can use these summary statistics:

$\sum x = 536$ $\sum x^2 = 30372$ $\sum y = 768$

$\sum y^2 = 54816$ $\sum xy = 40322$

3 Plot the data below on a scatter graph.

x	5.2	6.7	7.1	8.2	8.4	9.2	11.1
y	18.7	15.8	16.1	13.5	14	9.1	6.1

$\sum x = 55.9$ $\sum x^2 = 467.99$ $\sum y = 93.3$

$\sum y^2 = 1356.81$ $\sum xy = 697.14$

a Find the equation of the regression line of y on x.

b Plot the regression line directly onto your scatter graph.
Does it give a good line of best fit?

4 The variables P and S are known to be approximately linearly related. Six pairs of values, (p, q), of the variables gave the following results:

$\sum p = 150$ $\sum p^2 = 4450$ $\sum q = 145$

$\sum q^2 = 3656.2$ $\sum pq = 3950$

a Find the values of $\bar{p}$, $\bar{q}$, S_{pq}, S_{PP}.

b Find the equation of the regression line of q on p.

5 Plot the data below on a scatter graph.

x	4.1	6.3	7.5	8.7	8.1	10.3	13.5
y	19.3	17.2	14.1	12.5	15.5	11.2	6.1

$\Sigma x = 58.5$ $\Sigma x^2 = 542.39$ $\Sigma y = 95.9$

$\Sigma y^2 = 1426.29$ $\Sigma xy = 725.25$

a Find the equation of the regression line of y on x.

b Plot the regression line directly onto your scatter graph.
   Does it give a good line of best fit?

6 The variables P and S are known to be approximately linearly related.
   Eight pairs of values (p, q) of the variables gave the following results:

$\Sigma p = 103$ $\Sigma p^2 = 2531$ $\Sigma q = 85$

$\Sigma q^2 = 1582$ $\Sigma pq = 2163$

a Find the values of $\bar{p}, \bar{q}, S_{pq}, S_{pp}$.

b Find the equation of the regression line of q on p.

7 Plot the data below on a scatter graph.

x	13	17	22	28	31	33	43	51	62	69	75	78
y	31	56	38	51	59	49	82	67	72	95	81	95

Find the equation of the regression line of y on x.
You can use these summary statistics:

$\Sigma x = 522$ $\Sigma x^2 = 28\,540$ $\Sigma y = 776$
$\Sigma y^2 = 55\,032$ $\Sigma xy = 38\,512$

Consider again the mail order scenario introduced on page 120.

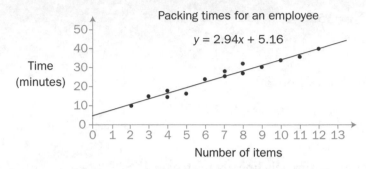

The equation of the regression line is $y = 2.94x + 5.16$.

So, for an order with ten items,
$$2.94 \times 10 + 5.16 = 34.56$$
The equation predicts that an employee will take roughly 35 minutes to make up the order.

This equation is calculated using the methods described on page 125.

This prediction is only an estimate – therefore 35 minutes is reasonably accurate.

Interpretation of the line

It is important to be able to interpret the gradient and y-intercept of the regression line in context.

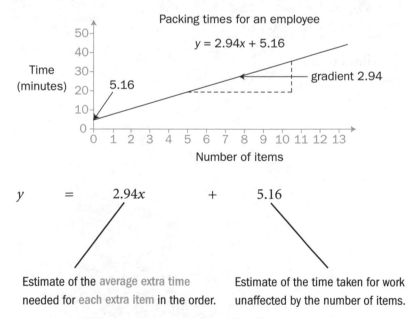

$$y \quad = \quad 2.94x \quad + \quad 5.16$$

Estimate of the average extra time needed for each extra item in the order.

Estimate of the time taken for work unaffected by the number of items.

Extrapolation and interpolation

The data for packing times considers orders up to 13 items.
Predicting for a number of items greater than 13 would involve
estimating beyond the range of the real data, and should be treated
with caution.

Prediction within the range of data is
called interpolation. As long as your
regression model is accurate it
should give reliable results.

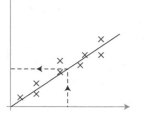

Prediction outside of the range of data
is called extrapolation. The further
beyond the range you go, the less
reliable the estimate.

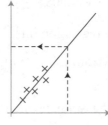

You should avoid extrapolation where possible. The relationship
between the variables may not match your initial assumptions.

Imagine you are exploring how much of a chemical is produced
in an experiment carried out at different temperatures.

You may find that a strong linear model
describes the situation well
over the range of your experiment.

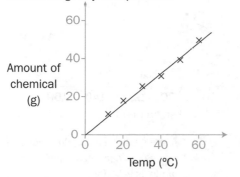

However if you increase the temperature
further, the substances may behave differently —
they could vaporise, explode, or reach a maximum.

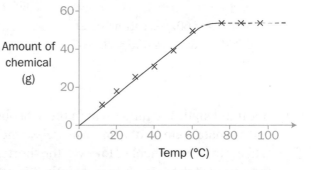

EXAMPLE 1

The table gives the lengths and widths of a random sample of leaves from a tree.

Length, x (mm)	86	89	91	103	107	116	117	125	125	127
Width, y (mm)	33	30	34	36	38	37	37	37	44	42

Length, x (mm)	127	134	144	145	146	149	149	151	152	155
Width, y (mm)	42	39	47	41	41	45	46	44	47	48

$\sum x = 2538$ $\sum x^2 = 331\,674$ $\sum y = 808$ $\sum y^2 = 33\,138$ $\sum xy = 104\,502$

a Calculate the equation of the regression line.

b Estimate the widths of leaves from this tree which have a length of
  i 120 mm   **ii** 60 mm

c How accurate do you think the estimates are in part **b**?

d A leaf from another tree has a length of 125 mm.
What width would you expect it to be?

First draw a scatter graph – this will help determine whether a linear regression model is reasonable.

a $y = 0.2048x + 14.406$

b **i** When $x = 120$, y is estimated at
    $0.2048 \times 120 + 14.406 = 38.982$
    or 40 mm.
  ii When $x = 60$, y is estimated at
    $0.2048 \times 60 + 14.406 = 26.694$
    or 27 mm.

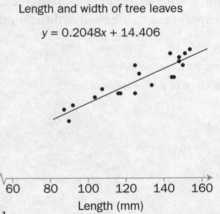

Length and width of tree leaves

$y = 0.2048x + 14.406$

c A leaf of length 120 mm is within the data observed, and the data seem to fit a linear model, so the estimate is likely to be reasonable. However, the shortest leaf in the observed data was 86 mm, so estimating the width of a leaf which is only 60 mm long requires extrapolation, and should be treated with caution.

A straight line fits the data quite well. So a linear regression model is valid for this data.

d Trying to use data from one context to make predictions about another is not a good idea. Here you do not know whether the trees are from the same species, or if they are similar ages or in similar condition – all of which might make a big difference to relationship – so you cannot really make any estimate from the information you have.

Remember that a line of regression is only an **estimate** of the underlying linear relation, and there are a number of things which affect the reliability of the predictions or estimates.

1 How close a fit are the data to a straight line? If the observed values lie close to a straight line it is reasonable to expect other data points to behave similarly. If they do not lie close, then your predictions are liable to more variation.

2 How many data points are used to obtain the line of regression? If the data set is small, then there is much more variation in the regression line than if a large data set is used.

3 Are you extrapolating outside the range of data used to produce the line of regression? If so, any predictions/estimated values need to have a 'health warning' attached.

Exercise 6.3

1 The scatter graph shows the age and the
   maximum distance at which a sample of
   drivers could read a motorway sign.

   The equation of the line of regression is
      $y = -0.8106x + 171.53$
   where y is the distance in metres and
   x is the age in years.

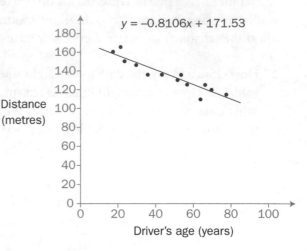

   a Estimate the maximum distance at which
      a driver of 60 years of age will be able to
      read the sign.

   b Mabel is 92 and still holds a driver's licence.
      Estimate the maximum distance at which she
      will be able to read the sign.

   c Comment on the reliability of your two estimates.

2 The scatter graph shows the times, in
   seconds, a number of Year 12 students took
   to complete a manual dexterity task with each
   of their left and right hands.

   The equation of the line of regression is
   $y = 0.8686x + 8.9918$ where y is the time
   with the left hand, and x is the time with
   the right hand.

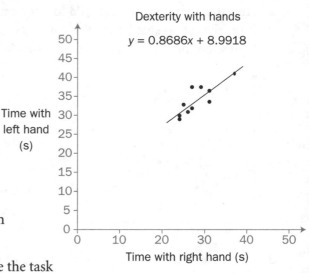

   a A student takes 27 seconds to complete the
      task with their right hand. Estimate the time
      the student will take to complete the task with
      their left hand.

   b Another student takes 47 seconds to complete the task
      with their right hand. Estimate the time this student
      will take to complete the task with their left hand.

   c Comment on the reliability of your two estimates.

3 The scatter graph shows the distances travelled, and the time taken, on a number of journeys a delivery van makes. The equation of the line of regression is $y = 1.6066x - 3.3799$ where y is the time in minutes and x is the distance in km.

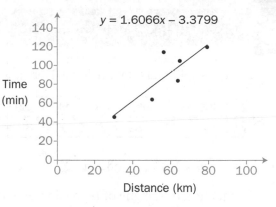

$y = 1.6066x - 3.3799$

a The delivery van has a journey of 62 km to make.
Estimate how long the journey will take.

b Another journey is only 2 km. What would the regression line predict for the journey time?

c Explain why you would not have wanted to use the regression line to predict the time for the journey in part **b**.

4 Anji has read that eating chocolate helps your short-term memory. She decides to carry out a simple memory test to see if there is a difference.
She shows her friends a grid which has 30 symbols on it. Two minutes later she asks them to write down as many of the symbols as they can remember.
She gives them each a bar of chocolate to eat, and half an hour later gives them a similar task, with a different set of 30 symbols.
She records the numbers of correct symbols for before, x, and after, y, eating the chocolate.

x	12	16	21	15	16	18	22	19
y	15	20	25	18	19	22	26	23

$\sum x = 139$ $\sum x^2 = 2491$ $\sum y = 168$ $\sum y^2 = 3624$ $\sum xy = 3004$

a Find the equation of the line of regression, y on x.

b Two of Anji's friends were away when she did this test, and wanted to try it. One of them remembered 17 of the symbols correctly on the first test. Estimate how many she will remember correctly on the second test.

c The other friend remembered 29 of the symbols correctly.
i Calculate how many symbols the regression line predicts she will remember in the second test.
ii Give two reasons why it is not a good idea to use the regression line for prediction in this case.

S1

You can simplify data by coding it. Here is a set of bivariate data.

x	400	410	420	430	440	450	460	470	480	490
y	82	83	89	87	90	89	93	95	99	105

Coding is less of an issue now with so much computing power available, but it is still an important technique.

$\sum x = 4450$ $\sum x^2 = 1\,988\,500$ $\sum y = 912$ $\sum y^2 = 83\,624$ $\sum xy = 407\,670$

The equation of the line of regression of y on x is $y = 0.222x - 7.51$

These data can be coded using $X = \dfrac{x-400}{10}$, $Y = y - 80$.

This gives a new data set with much simpler sums.

This coding makes the numbers easier to work with.

X	0	1	2	3	4	5	6	7	8	9
Y	2	3	9	7	10	9	13	15	19	25

$\sum X = 45$ $\sum X^2 = 285$ $\sum Y = 112$ $\sum Y^2 = 1704$ $\sum XY = 687$

The equation of the line of regression of Y on X is $Y = 2.22X + 1.22$.

You can work with the coded data and then obtain the regression line in the original variables by substitution and then simplifying the resulting equation:

$$y - 80 = 2.22\left(\frac{x-400}{10}\right) + 1.22, \text{ so } y = 0.222x - 7.58$$

EXAMPLE 1

The journeys taken by the delivery van in question 3 on page 133 have distances recorded in km and times in minutes. The driver is paid for his work by the number of miles driven and his time in hours, so the data is coded:

   5 miles = 8 km, 60 minutes = 1 hour

Kilometres	x	32	48	57	64	65	79
Minutes	y	45	65	115	84	105	120

Miles	X	20	30	35.6	40	40.6	49.4
Hours	Y	0.75	1.1	1.9	1.4	1.75	2

The regression line of the coded data is $Y = 0.0428X - 0.056$. Find the regression line of y on x for the distance in km and the time in minutes.

$Y = 0.0428X - 0.056$

$X = x \times \dfrac{5}{8}$, $Y = \dfrac{y}{60}$

$\therefore \dfrac{y}{60} = 0.0428\left(\dfrac{5x}{8}\right) - 0.058$

$\therefore y = 1.61x - 3.48$

S1

Exercise 6.4

1 **a** Code the data set given in the table by $X = x - 200$ and $Y = y - 130$.

x	207	203	214	208	216	213	203	205
y	132	134	145	135	147	141	128	131

  b Find the line of regression of Y on X.

  c Find the line of regression of y on x from your answer to part **b**.

2 A set of data is coded using $x = p - 300$ and $y = \frac{q}{5}$.

$\sum x = 136$ $\sum x^2 = 3074$

$\sum y = 41$ $\sum y^2 = 251$

$\sum xy = 733$ $n = 10$

  a Find the line of regression of y on x.

  b Find the line of regression of q on p.

3 The manager of an athletics club finds some old club records of long jump competitions in which the winning jump d (feet) and the temperature F (°F) are recorded. He converts six of the results into the units used in today's competitions, D(metres) and C(°C), by $C = \frac{5}{9} \times (F - 32)$; $D = \frac{d}{3.3}$

$\sum C = 146.1$ $\sum C^2 = 3656.31$

$\sum D = 38.53$ $\sum D^2 = 248.2$

$\sum CD = 946.6$

  a Find the regression line of D on C.

  b Find the regression line of d on F.

  c Estimate the length of the winning jump in feet on a day when the temperature was 71°F.

1 a An experiment was conducted to see whether there was any relationship between the maximum tidal current, y cm s^{-1}, and the tidal range, x metres, at a particular marine location. (The *tidal range* is the difference between the height of high tide, and the height of low tide.) Readings were taken over a period of ten days, and the results are shown in the following table.

X	2.1	2.5	3.1	3.0	3.5	3.8	4.1	4.4	4.8	5.0
Y	14.5	21.4	24.9	31.5	34.7	41.3	16.4	49.8	58.3	60.3

a Draw a scatter diagram for the data.

b Calculate exact values of S_{xy} and S_{xx}.

c Calculate the equation for the regression line of maximum tidal current on tidal range. Draw the regression line on your scatter diagram.

d Estimate the maximum tidal current on a day when the tidal range is 3.8 metres. Comment on the reliability of your estimate.

e Explain why the equation in part **c** should not be used to predict the tidal range when the maximum tidal current is 37 cm s^{-1}.

[(c) Edexcel Limited specimen]

2 A long distance lorry driver recorded the distance travelled, d km, and the time taken t hours, each day. Summarised below are data from the driver's records for a random sample of nine days.

The data are coded such that $x = \dfrac{(d - 300)}{10}$ and $y = t - 5$.

$$\sum x = 40.5 \quad \sum y = 9.5 \quad \sum xy = 49.4 \quad S_{xx} = 43.04$$

a Find the equation of the regression line of y on x in the form $y = a + bx$.

b Hence find the equation of the regression line of t on d.

c Predict the time taken on a journey of 340 km.

3 A psychologist believes that people's ability to cope with stress depends on the stability of their home background. Eight students took a test to assess their ability to cope with stress and filled out a questionnaire about their home background. The scores for each student are given in the table, where s represents the score on the stress test and r the score the psychologist gave to the responses to the questionnaire on home background.

Mark		Student							
		A	B	C	D	E	F	G	H
	s	18	23	25	19	13	17	27	16
	r	11	17	17	13	9	12	16	10

a   Write down which is the explanatory variable in this investigation.

b   Draw a scatter diagram to illustrate these data.

c   Showing your working, find the equation of the appropriate regression line.

d   Draw the regression line on your scatter diagram.

A ninth student was absent for the questionnaire, but she sat the stress test and scored 23.

e   Can this model be used to estimate the score she would have had on the questionnaire about home stability? You must give a reason.

4 The heating in an office is switched on at 7.00 am each morning. On a particular day, the temperature of the office, t °C, was recorded m minutes after 7.00 am. The results are shown in the table.

m	0	10	20	30	40	50
t	6.0	8.9	11.8	13.5	15.3	16.1

a   Calculate the exact values of S_{mt} and S_{mm}.

b   Calculate the equation of the regression line of t on m in the form $t = a + bm$.

c   Use your equation to estimate the value of t at 7.35 am

d   State, giving a reason, whether or not you would use the regression equation in part b to estimate the temperature

   i   at 9.00 am that day
   ii   at 7.15 am one month later.   [(c) Edexcel Limited 2004]

S1

5 A long distance lorry driver recorded the distance travelled,
m miles, and the amount of fuel used, f litres, each day.
Summarised below are data from the driver's records for
a random sample of eight days.
The data are coded such that $x = m - 250$ and $y = f - 100$.

$$\sum x = 130 \quad \sum y = 48 \quad \sum xy = 8880 \quad \sum x^2 = 20\,487.5$$

  a Find the equation of the regression line of y on x in the
  form $y = a + bx$.

  b Hence find the equation of the regression line of f on m. [(c) Edexcel Limited 2005]

6 The following table shows the height, x, to the nearest cm,
and the weight, y, to the nearest kg, of a random sample of
12 students.

x	148	164	156	172	147	184	162	155	182	165	175	152
y	39	59	56	77	44	77	65	49	80	72	70	52

  a On graph paper, draw a scatter diagram to represent these data.

  b Write down, with a reason, whether the correlation coefficient
  between x and y is positive or negative.

The data in the table can be summarised as follows.

$$\sum x = 1962 \quad \sum y = 740 \quad \sum y^2 = 47\,746 \quad \sum xy = 122\,783 \quad S_{xx} = 1745$$

  c Find S_{xy}.

The equation of the regression line of y on x is $y = -106.331 + bx$.

  d Find, to three decimal places, the value of b.

  e Find, to three significant figures, the mean, $\bar{y}$, and the standard
  deviation, s, of the weights of this sample of students. [(c) Edexcel Limited 2005]

7 The bulb of an industrial lamp can operate at different voltages,
but in general the higher the voltage the sooner the bulb needs
to be replaced. Eleven pairs of observations of voltage, v volts,
and life of bulb, t hours, are collected.
For convenience the data are coded so that $x = \dfrac{(v - 110)}{10}$ and
$y = t - 90$ and the following summations obtained.

$$\sum x = 66 \quad \sum y = 506 \quad \sum x^2 = 354.5 \quad \sum y^2 = 13\,723.6 \quad \sum xy = 1270.1$$

  a Find the equation of the regression line of t on v.

  b Interpret the slope of your regression line.

  c Estimate the life of a bulb operated at 170 volts.

8 Ms Darnton advises students on university entrance.
She has GCSE and GCE results from students in the
previous year.
At GCSE she takes only the best eight results and awards
8 points for an A*, 7 for an A, etc.
At GCE she awards 5 for an A at AS-level down to 1 for grade E,
and double for grades at A-level, i.e. 10 for a grade A down to
2 for a grade E.

The results for GCSE, x, and GCE, y, are shown below for a
sample of the students.

x	62	53	56	55	64	57	48	53	62	55	59	45
y	35	32	30	28	40	31	23	18	33	36	29	19

a On graph paper, draw a scatter diagram to represent
these data.

The data in the table can be summarised as follows.

$$\sum x = 669 \quad \sum y = 354 \quad \sum x^2 = 37\,647 \quad \sum y^2 = 10\,934 \quad \sum xy = 20\,063$$

b Calculate the correlation coefficient between x and y.

c Find the equation of the regression line of y on x.

d Alisha had one grade A*, one A, three Bs, three Cs and one D in
her GCSEs and is thinking of applying for a course at a university
which usually asks for two As and a B at A-level with a B at AS.
What advice should Ms Darnton give Alisha?

e Karin is a pupil at another school. She had one A, one B, four Cs,
a D and an F in her GCSEs. Since she knows Ms Darnton,
Karin asks her what grades she might get if she went on to
take A-levels.
Give two reasons why Ms Darnton should not use the
regression line calculated in part **c** to predict Karin's results.

Summary

Refer to

○ The line of regression is the best fit by the *least squares* criterion, so it is the line which gives the lowest sum of squared differences between the observed values of the dependent variable and the values predicted by the line.

6.1–6.2

○ $(\bar{x}, \bar{y})$ always lies on any line of regression. The equation of the line y on x is given by $y - \bar{y} = b(x - \bar{x})$

where $b = \dfrac{\sum(x_i - \bar{x})(y_i - \bar{y})}{\sum(x_i - \bar{x})^2} = \dfrac{\sum x_i y_i - \dfrac{(\sum x_i)(\sum y_i)}{n}}{\sum x_i^2 - \dfrac{(\sum x_i)^2}{n}} = \dfrac{S_{xy}}{S_{xx}}$

6.1–6.2

○ An estimate or prediction of the value of y can be made for a value of x by substituting the value of x into the equation of the line of regression. If the value of x is outside the range of observed values of x used in constructing the regression line, this is known as extrapolation, and needs to be treated with caution.

6.3

○ If the bivariate data (s, t) is coded using $x = \dfrac{s - k}{p}$, $y = \dfrac{t - l}{q}$ and the regression line $y = a + bx$ is found then the regression line of the original data is $\dfrac{t - l}{q} = a + b\left(\dfrac{s - k}{p}\right)$

6.4

Links

Regression modelling is used widely in business to make predictions from existing data about the likely behaviour in situations where a simple linear model provides a good description of the behaviour.

The widespread use of computing now allows more sophisticated modelling where non-linear functions, or more than one explanatory variable, are introduced into the model being fitted to the data.

S1

Discrete random variables

This chapter will show you how to
- find missing values in probability distributions
- find the probability distribution explicitly given a probability function
- move between the probability distribution and the cumulative distribution function
- calculate the mean, variance and standard deviation of a discrete random variable
- calculate the mean of functions of a random variable including a linear change of scale
- work with the discrete uniform distribution.

Before you start

You should know how to:

1 Solve linear simultaneous equations.

e.g. Solve the simultaneous equations

$$x + 3y = 9 \quad (1)$$
$$2x - y = 4 \quad (2)$$

Multiply (1) by 2:
$$2x + 6y = 18 \quad (3)$$

Eqn (3) − eqn (2):
$$7y = 14$$
$$y = 2$$

Substitute 2 for y in (1)
$$x + 6 = 9$$
$$x = 3$$

2 Substitute into simple expressions.

e.g Evaluate the expression
$$\frac{1}{x} + \frac{1}{(x+1)} + \frac{1}{(x+2)} \text{ if } x = 2.$$

$$\frac{1}{2} + \frac{1}{(2+1)} + \frac{1}{(2+2)}$$

$$= \frac{1}{2} + \frac{1}{3} + \frac{1}{4} = \frac{13}{12} = 1\frac{1}{12}$$

Check in

1 Solve the simultaneous equations
$$0.6 + a + b = 1$$
$$3.2 + 2a + 4b = 4.6$$

2 Write down the values of $\frac{12}{x}$ for $x = 1, 2, 3$ and 4.

Discrete random variables

A **random variable** is a quantity that can take any value determined by the outcome of a random event.

Random variables can arise from probability experiments. When you throw two dice,
X, the sum of scores showing, is a random variable.
Similarly,
Y, the product of the scores, and Z, the larger of the two scores, are also random variables.

Random variables can arise from real-life observation, for example:

X, the number of telephone calls arriving at a switchboard between 10.00 and 10.30 a.m.

When values of a variable have a probability attached, they form a **probability distribution**.
If a random variable, X, can only take distinct values, it is a **discrete random variable**.

Similarly, when values have a frequency attached, they form a **frequency distribution**.

> o X is a **discrete random variable** if X takes values x_1, x_2, x_3, ... and $P(X = x_i) = p_i$ where all $p_i \geqslant 0$ and $\sum p_i = 1$.

$P(X = x_1) = p_1$ just means that each value of X has a probability attached to it.

Note that this means that X can only take distinct values, and the sum of all the probabilities of X is 1.

EXAMPLE 1

If X is the sum of the scores on two dice, the sample space for X will be:

The sample space is just a list of all possible outcomes.

X	1	2	3	4	5	6
1	2	3	4	5	6	7
2	3	4	5	6	7	8
3	4	5	6	7	8	9
4	5	6	7	8	9	10
5	6	7	8	9	10	11
6	7	8	9	10	11	12

Find the probability distribution of X.

The probability distribution of X is:

x	2	3	4	5	6	7	8	9	10	11	12
$P(X = x)$	$\frac{1}{36}$	$\frac{2}{36}$	$\frac{3}{36}$	$\frac{4}{36}$	$\frac{5}{36}$	$\frac{6}{36}$	$\frac{5}{36}$	$\frac{4}{36}$	$\frac{3}{36}$	$\frac{2}{36}$	$\frac{1}{36}$

Notice that the sum of the probabilities is 1.

EXAMPLE 2

X is a random variable with probability distribution given by

x	-2	-1	0	1	2
$P(X = x)$	0.1	0.1	0.4	a	0.1

Find the value of a.

A probability distribution can be in the form of a table.

$\Sigma\, p_i = 1$

$1 - (0.1 + 0.1 + 0.4 + 0.1) = 0.3$

$$\therefore a = 0.3$$

Exercise 7.1

1 Which of the following are discrete random variables?
Give a reason for any which are not.

a

x	1	2	3	4	5
$P(X = x)$	0.2	0.3	0.4	0.3	0.2

b

x	-2	-1	0	1	2
$P(X = x)$	0.2	0.3	0.1	0.3	0.1

c

x	5	7	10	15	20
$P(X = x)$	$\frac{1}{2}$	$\frac{1}{4}$	$\frac{1}{8}$	$\frac{1}{16}$	$\frac{1}{16}$

d

x	1	2	3	4	5
$P(X = x)$	0.2	0.3	0.4	0.3	-0.2

2 A fair die is thrown. List the probability distribution for the
following random variables:

a X = score on the die

b $Y = 2 \times$ score on the die

c Z = the square of the score on the die

d $W = 0$ if the score on the die is a factor of 6; 1 otherwise.

3 Two coins are tossed. X is the number of heads seen. List the
probability distribution for X.

4 a

x	1	2	3	4	5
$P(X = x)$	0.2	0.1	0.3	a	0.1

i Find a. **ii** Find $P(X \geqslant 2)$.

b

x	-2	-1	0	1
$P(X = x)$	k	$2k$	$2k$	k

i Find k. **ii** Find $P(X \leqslant 0)$.

c

x	5	6	8	9	12
$P(X = x)$	0.4	0.2	0.3	a	0.1

i Find a. **ii** Find $P(X \geqslant 9)$.

The probability function *p(x)*

You can write the probability distribution of a discrete random variable:

- as a **possibility space** – a table of values with their associated probabilities
- as a **probability function** – a formula for $p(x)$.

Here is an example of a probability function:

$P\{X = r\} = kr \quad r = 1, 2, 3, 4$

This means that the random variable X can take the values 1, 2, 3 or 4 with probabilities $k, 2k, 3k$ and $4k$.

> The total probability must be 1: when you add these up you get $10k$, so $k = 0.1$.

You can then draw up a table showing all the probabilities explicitly:

r	1	2	3	4
Probability, $P(X = r)$	0.1	0.2	0.3	0.4

EXAMPLE 1

$$P\{X = r\} = \left(\frac{1}{6}\right)\left(\frac{5}{6}\right)^{r-1} \quad r = 1, 2, 3, 4, \dots$$

a Write down the probabilities that $X = 1$, $X = 2$ and $X = 3$.

b Find the probability that $X \geqslant 4$.

---

a $P\{X = 1\} = \left(\frac{1}{6}\right)\left(\frac{5}{6}\right)^{0} = \frac{1}{6}$

$P\{X = 2\} = \left(\frac{1}{6}\right)\left(\frac{5}{6}\right)^{1} = \frac{5}{36}$

$P\{X = 3\} = \left(\frac{1}{6}\right)\left(\frac{5}{6}\right)^{2} = \frac{25}{216}$

$P\{X \geqslant 4\} = 1 - P\{X = 1, 2, 3\} = 1 - \left(\frac{1}{6} + \frac{5}{36} + \frac{25}{216}\right) = \frac{125}{216}$

EXAMPLE 2

The random variable, X, has probability distribution

$$P\{X = k\} = \begin{cases} \dfrac{k-1}{36} & k = 2, 3, 4, 5, 6, 7 \\[2mm] \dfrac{13-k}{36} & k = 8, 9, 10, 11, 12 \end{cases}$$

> This probability function describes the sum of scores on two fair dice.

Find   **a** $P(X = 3)$ **b** $P(X > 9)$

---

a $P(X = 3) = \frac{3-1}{36} = \frac{2}{36} = \frac{1}{18}$

b $P(X > 9) = P(X = 10, 11 \text{ or } 12) = \frac{3}{36} + \frac{2}{36} + \frac{1}{36} = \frac{6}{36} = \frac{1}{6}$

Exercise 7.2

1 Which of the following could not be the probability distribution of a random variable? For those which could be, find the probability distribution.

  a $P\{X = r\} = kr$ $r = 1, 2, 3, 4, 5$

  b $P\{X = r\} = \dfrac{k}{r}$ $r = 1, 2, 3, 4$

  c $P\{X = r\} = \dfrac{k}{r-1}$ $r = 1, 2, 3, 4$

  d $P\{X = r\} = \dfrac{k}{r+1}$ $r = 1, 2, 3, 4$

2 For each of the following probability distribution functions, list the probability distribution.

  a $P\{Z = z\} = \dfrac{5-z}{10}$ $z = 1, 2, 3, 4$

  b $P\{Y = y\} = \dfrac{1}{5}$ $y = 1, 2, 3, 4, 5$

  c $P\{W = w\} = k(w-1)$ $w = 2, 3, 4, 5, 6, 7$
      $\phantom{P\{W = w\}} = k(13 - w)$ $w = 8, 9, 10, 11, 12$

  d $P\{H = r\} = k(2r-1)$ $r = 1, 2, 3, 4, 5, 6$

> In part **c**, the formula changes from $w = 8$ onwards.

3 X is a discrete random variable with probability distribution

  $P\{X = x\} = \dfrac{x}{15}$ $x = 1, 2, 3, 4, 5$

  A is the event $X \geqslant 3$ and B is the event $X < 4$.

  Find

  a $P(A)$ b $P(B)$ c $P(A \cap B)$

  d $P(A \mid B)$ e $P(B \mid A)$.

4 Z is a discrete random variable with probability distribution

  $P\{Z = z\} = \dfrac{k}{z}$ $z = 1, 2, 3, 4$

  a Show that $k = 0.48$.

  b Find   i $P(Z > 1)$ ii $P(Z = 4 \mid Z > 1)$.

5 $P\{Y = y\} = cy^2$ for $y = 1, 2, 3, 4$

  a Find the value of c.

  b Hence find $P\{Y < 3\}$.

$F(x) = P(X \leqslant x)$ is the **cumulative distribution function** for the random variable X.

You have probably met cumulative frequency before. $F(x)$ is cumulative probability.

If two dice are thrown and the higher of the two scores is recorded, the possibility space looks like this:

	1	2	3	4	5	6
1	1	2	3	4	5	6
2	2	2	3	4	5	6
3	3	3	3	4	5	6
4	4	4	4	4	5	6
5	5	5	5	5	5	6
6	6	6	6	6	6	6

$$F(3) = P(X \leqslant 3) = \frac{9}{36}, \quad F(5) = P(X \leqslant 5) = \frac{25}{36}$$

The cumulative distribution function can be expressed as

$$F(x) = P(X \leqslant x) = \frac{x^2}{36}$$

S1

EXAMPLE 1

The random variable, X, has probability distribution

x	15	30	40
p	0.5	0.3	0.2

Compile a table for the cumulative distribution function.

The cumulative distribution function (cdf) may be shown in the form of a table.

The cumulative distribution function for X is

x	15	30	40
F(x)	0.5	0.8	1

EXAMPLE 2

The random variable, X, has cumulative distribution function given by the table

x	-3	-2	-1	0	1	2	3
F(x)	0.2	0.3	0.5	0.6	0.75	0.9	1

Find $P(X = 2)$ and $P(|X| < 2)$.

$P(X = 2) = F(2) - F(1) = 0.9 - 0.75 = 0.15$

$P(|X| < 2) = P(X = -1 \text{ or } 0 \text{ or } 1)$
$= F(1) - F(-2) = 0.75 - 0.3 = 0.45$

EXAMPLE 3

The random variable, X, has cumulative distribution in the form

$$F(x) = \frac{(x + a)^2}{25} \quad x = 1, 2, 3; \ a > 0$$

Find the value of a, and give the probability distribution of X.

Since 3 is the largest value of x with a cumulative probability,

$F(3) = 1$, $\frac{(3 + a)^2}{25} = 1$, so $a = 2$

Then $F(1) = \frac{9}{25}$; $F(2) = \frac{16}{25}$ and $F(3) = 1$.

So the probability distribution is

x	1	2	3
p	$\frac{9}{25}$	$\frac{7}{25}$	$\frac{9}{25}$

$p(2) = F(2) - F(1)$

Exercise 7.3

1 The random variable, X, has probability distribution

x	1	2	3	4
p	0.3	0.2	0.3	0.2

a Find the cumulative distribution function $F(X)$.

b Write down the value of

  i $F(0)$ **ii** $F(6)$ **iii** $F(2.5)$.

2 For the following probability distributions, write down the cumulative distribution function.

a

x	1	2	3	4	5
$P(X = x)$	0.2	0.1	0.2	0.3	0.2

b

X	5	7	10	15	20
$P(X = x)$	$\frac{1}{2}$	$\frac{1}{4}$	$\frac{1}{8}$	$\frac{1}{16}$	$\frac{1}{16}$

3 For the following the cumulative distribution functions, write down the probability distribution.

a

x	-2	-1	0	1	2
$F(x)$	0.1	0.2	0.4	0.7	1

b

x	5	7	10	15	20
$F(x)$	$\frac{1}{4}$	$\frac{1}{3}$	$\frac{1}{2}$	$\frac{3}{4}$	1

S1

The **mean** or **expected value** of a probability distribution is defined as

$$\mu = E(X) = \Sigma\, px$$

This means that you:

- multiply each score by its probability
- sum over all possible values of the random variable.

EXAMPLE 1

Members of a public library may borrow up to five books at any one time.
The number of books borrowed by a member on each visit is a random variable, X, with the following probability distribution.

X	0	1	2	3	4	5
Probability	0.24	0.12	0.14	0.30	0.05	0.15

Find the mean of X.

$$E(X) = 0 \times 0.24 + 1 \times 0.12 + 2 \times 0.14 + 3 \times 0.30$$
$$+ 4 \times 0.05 + 5 \times 0.15$$
$$= 2.25$$

The expected value does not have to be a possible outcome.

The next example uses the fact that the sum of all the probabilities is 1.

EXAMPLE 2

X is a discrete random variable with probability distribution

x	1	2	3	4
p	5k	2k	k	2k

Don't get confused over X and x: X can take a range of values, x is one particular value.

Show that $k = 0.1$ and calculate $E(X)$.

$5k + 2k + k + 2k = 10k = 1,$ so $k = 0.1$

x	1	2	3	4
p	0.5	0.2	0.1	0.2

$$E(X) = 1 \times 0.5 + 2 \times 0.2 + 3 \times 0.1 + 4 \times 0.2 = 2.0$$

To calculate the mean of a function, $g(X)$, of a random variable, you can replace the value of X by the value of $g(X)$, keeping the same probability distribution.

So to find the mean of $Y = 2X + 5$ you would use (for the library books in Example 1):

$Y = 2X + 5$	5	7	9	11	13	15
Probability	0.24	0.12	0.14	0.30	0.05	0.15

$$E(Y) = 5 \times 0.24 + 7 \times 0.12 + 9 \times 0.14 + 11 \times 0.30 + 13 \times 0.05$$
$$+ 15 \times 0.15 = 9.5$$

You will meet another way to calculate this in Section 7.6

To find $E(X^2)$, you can list the values of X^2 with the probability distribution of X.

So for the library books example:

X	0	1	4	9	16	25
Probability	0.24	0.12	0.14	0.30	0.05	0.15

$$E(X^2) = 0 \times 0.24 + 1 \times 0.12 + 4 \times 0.14 + 9 \times 0.30 + 16 \times 0.05$$
$$+ 25 \times 0.15$$
$$= 7.93$$

$E(X^2)$ will become useful in the next section.

EXAMPLE 3

X is a discrete random variable with probability distribution

x	1	2	3	4
p	a	0.3	b	0.2

If $E(X) = 2.7$ find the values of a and b.

$a + 0.3 + b + 0.2 = 1$

$a + b = 0.5$

$E(X) = 1 \times a + 2 \times 0.3 + 3 \times b + 4 \times 0.2$

$\qquad = a + 0.6 + 3b + 0.8$

$\qquad = a + 3b + 1.4$

So $\quad a + 3b + 1.4 = 2.7$

$\qquad\qquad a + 3b = 1.3$

$\qquad\qquad\quad 2b = 0.8$

so $b = 0.4$ and $a = 0.1$

Exercise 7.4

1 For the following probability distributions, calculate the expectation of X.

a

x	1	2	3	4	5
$P(X = x)$	0.2	0.1	0.2	0.3	0.2

b

x	-2	-1	0	1	2
$P(X = x)$	0.2	0.3	0.1	0.3	0.1

c

x	5	7	10	15	20
$P(X = x)$	$\frac{1}{2}$	$\frac{1}{4}$	$\frac{1}{8}$	$\frac{1}{16}$	$\frac{1}{16}$

2 For each of the following probability distribution functions calculate the mean.

a $P\{Z = z\} = \dfrac{5 - z}{10}$ $z = 1, 2, 3, 4$

b $P\{Y = y\} = \dfrac{1}{5}$ $y = 1, 2, 3, 4, 5$

3 a

x	1	2	3	4	5
$P(X = x)$	0.3	0.2	0.1	a	0.1

　　i　Find a.
　　ii　Find $E(X)$.
　　iii　Find $E(X^2)$.

b

x	5	6	8	12
$P(X = x)$	k	$2k$	$2k$	k

　　i　Find k.
　　ii　Find $E(X)$.
　　iii　Find $E(X^2)$.

4 a

x	1	2	3	4	5
P(X = x)	0.1	0.3	0.2	0.2	0.2

Calculate   **i** E(X)
                 ii E(3X − 2).

b

x	-2	-1	0	1	2
P(X = x)	0.3	0.4	0.1	0.1	0.1

Calculate   **i** E(X)
                 ii E(4X + 6).

c

x	1	2	3	4	5
P(X = x)	$\frac{1}{2}$	$\frac{1}{4}$	$\frac{1}{8}$	$\frac{1}{16}$	$\frac{1}{16}$

Calculate   **i** E(X)
                 ii E(25 − X).

5 Y is the larger score showing when two dice are thrown.
Calculate E(Y).

6 X is a probability distribution function and E(X) = 3.4.

x	1	2	3	4	5
P(X = x)	0.1	0.2	0.1	a	b

Find a and b.

7 X is a probability distribution function and E(X) = 7.4.

x	4	6	7	10	11
P(X = x)	0.2	a	0.3	b	0.2

a Find a and b.

b Find the probability that X > E(X).

The **variance** of a probability distribution is defined as
$$\text{Var}(X) = E[\{X - E(X)\}^2]$$
The alternative version (easier to use in practice) is:
$$\text{Var}(X) = E(X^2) - \{E(X)\}^2$$

$E(X^2) = \sum px^2$
You multiply all the x^2 values by their probabilities, and then add up the total.

An easy way to recall the alternative version is
'Find the mean of the squares then subtract the square of the mean.'

EXAMPLE 1

X is a random variable with probability distribution

x	1	2	3	4
$P\{X = x\}$	a	0.2	$3a$	0.2

a   Find the value of a.   **b**   Calculate $E(X)$ and $\text{Var}(X)$.

---

a   Since $\Sigma p = 1$, $4a + 0.4 = 1$ so $a = 0.15$

b   In table form:

x	p	px	px^2
1	0.15	0.15	0.15
2	0.2	0.4	0.8
3	0.45	1.35	4.05
4	0.2	0.8	3.2
		$\Sigma px = 2.7$	$\Sigma px^2 = 8.2$

$E(X) = \Sigma px = 2.7$
$\text{Var}(X) = E(X^2) - \{E(X)\}^2$
$\phantom{\text{Var}(X)} = 8.2 - 2.7^2$
$\phantom{\text{Var}(X)} = 0.91$

EXAMPLE 2

X is a random variable with probability distribution

x	1	2	3	4
$P\{X = x\}$	a	0.2	b	0.2

If $E(X) = 2.6$

a   find the values of a and b **b**   calculate $E(X^2)$ and $\text{Var}(X)$.

---

a $a + b + 0.4 = 1$
$a + 0.4 + 3b + 0.8 = 2.6$, so $a + 3b = 1.4$
Solving gives $a = 0.2$, $b = 0.4$

b   In table form:

x	p	px^2
1	0.2	0.2
2	0.2	0.8
3	0.4	3.6
4	0.2	3.2
		$\Sigma px^2 = 7.8$

$\text{Var}(X) = E(X^2) - \{E(X)\}^2$
$\phantom{\text{Var}(X)} = 7.8 - 2.6^2$
$\phantom{\text{Var}(X)} = 1.04$

S1

Exercise 7.5

1 For each probability distribution, calculate $E(X)$ and $Var(X)$.

a

x	5	6	7	8	9
$P(X = x)$	0.1	0.2	0.3	0.3	0.1

b

x	-2	-1	0	1	2
$P(X = x)$	0.1	0.2	0.3	0.3	0.1

2 For each probability distribution function, calculate the mean and variance.

a $P\{Z = z\} = \dfrac{6 - z}{15}$ $z = 1, 2, 3, 4, 5$

b $P\{Y = y\} = \dfrac{1}{6}$ $y = 1, 2, 3, 4, 5, 6$

c $P\{W = w\} = k(w - 1)$ $w = 2, 3, 4, 5, 6, 7$
$\qquad\qquad\quad = k(13 - w)$ $w = 8, 9, 10, 11, 12$

3 The random variable, X, has probability distribution

a

x	1	2	3	4	5
$P(X = x)$	0.2	0.2	0.2	a	0.1

    i Find a.   **ii** Find $E(X)$ and $Var(X)$.

b

x	3	4	5	6
$P(X = x)$	k	$2k$	$2k$	k

    i Find k.   **ii** Find $E(X)$ and $Var(X)$.

4 Y is the smaller score showing when two dice are thrown. Calculate $E(Y)$ and $Var(Y)$.

5 X is a probability distribution function and $E(X) = 3.7$.

x	1	2	3	4	5
$P(X = x)$	0.1	0.2	0.1	a	b

Find a, b and $Var(X)$.

6 X is a probability distribution function and $E(X) = 5.7$.

x	1	2	4	8	16
$P(X = x)$	0.1	a	0.3	b	0.1

Find a, b and $Var(X)$.

S1

You can find the expectation of any function of a discrete random variable:

$$E\{f(X)\} = \Sigma[f(x) \times P(X = x)]$$

Where the function f(X) is linear, these properties are always true:

$$E(aX + b) = aE(X) + b$$
$$Var(aX + b) = a^2Var(X)$$

You already know that
$E(X^2) = \Sigma\,px^2$

The standard deviation of $aX + b$ will then be $|a|$ times the standard deviation of X.

EXAMPLE 1

a   E(X) = 3 and Var(X) = 2.1.
Find the mean and variance of 2X + 5.

b   E(X) = 7 and Var(X) = 5.3.
Find the mean and variance of 4 − X.

c   E(X) = 4 and Var(X) = 3.
Find E(X^2) and E{(X + 2) (2X − 3)}.

---

a   E(2X + 5) = 2 × 3 + 5 = 11
Var(2X + 5) = 4 × 2.1 = 8.4

b   E(4 − X) = 4 − 7 = −3
Var(4 − X) = (−1)^2 × 5.3 = 5.3

E(4 − X) = E(−X + 4)
so a = −1 and b = 4

c   Var(X) = E(X^2) − {E(X)}^{2} so 3 = E(X^2) − 16 ∴ E(X^2) = 19
E{(X + 2)(2X − 3)} = E{2X^2 + X − 6} = 2E(X^2) + E(X) − 6
                                              = 2 × 19 + 4 − 6 = 36

Making comparisons between sets of data or random variables which have a different mean and spread is difficult, so applying a linear change of scale to make the mean and variance take specified values can be useful.

EXAMPLE 2

E(X) = 7 and Var(X) = 9.
Find a and b so that $Y = aX + b$ has mean 10 and variance 1 (a > 0).

E(aX + b) = aE(X) + b = 7a + b
Var(aX + b) = a^2 Var(X) = 9a^2

Need to find a, b so that 9a^2 = 1 and 7a + b = 10.

$a = \frac{1}{3}$ and $b = 10 - \frac{7}{3} = \frac{23}{3}$ so $Y = \frac{X + 23}{3}$

This method is known as **standardising** the data.

SI

EXAMPLE 3

The probability distribution of the number of teachers who need cover in a school at a particular time of year is given by

x	1	2	3	4
$P(X=x)$	0.3	0.4	0.2	0.1

a Find the mean and standard deviation of X.

The cost of providing cover is £185 per teacher plus an administrative fee of £30.

b Find the mean and standard deviation of the cost of providing cover.

- - - - -

a $E(X) = 1 \times 0.3 + 2 \times 0.4 + 3 \times 0.2 + 4 \times 0.1 = 2.1$
$E(X^2) = 1^2 \times 0.3 + 2^2 \times 0.4 + 3^2 \times 0.2 + 4^2 \times 0.1 = 5.3$
$Var(X) = 5.3 - 2.1^2 = 0.89$
standard deviation of $X = \sqrt{0.89} = 0.943\ldots$

b $C = 185X + 30$, so $E(C) = 185 \times 2.1 + 30 = £418.50$
standard deviation of $C = 185 \times 0.943\ldots = £174.53$

Exercise 7.6

1 $E(X) = 5.7, Var(X) = 1.9$

For each of the following functions, write down the mean and variance.

a $2X + 7$ **b** $4 - 3X$ **c** $X + 3$ **d** $7X$

2 a

x	1	2	3	4	5
$P(X=x)$	0.1	0.2	0.3	0.3	0.1

i Calculate $E(X)$ and $Var(X)$.
ii Calculate the mean and variance of $5X - 2$.

b

x	-2	-1	0	1	2
$P(X=x)$	0.1	0.4	0.2	0.2	0.1

i Calculate $E(X)$ and $Var(X)$.
ii Calculate the mean and variance of $7 - 2X$.

c

x	7	8	9	10	16
$P(X=x)$	$\frac{1}{2}$	$\frac{1}{4}$	$\frac{1}{8}$	$\frac{1}{16}$	$\frac{1}{16}$

i Calculate $E(X)$ and $Var(X)$.
ii Calculate the mean and variance of $4 + 3X$.

The scores on a fair die provide six possible outcomes, 1, 2, 3, 4, 5 and 6, each with probability $\frac{1}{6}$.

This is an example of the **discrete uniform distribution**.

X follows a discrete uniform distribution if X takes any of the values 1, 2, 3, ..., n each with probability $\frac{1}{n}$.

The mean $E(X) = \frac{n+1}{2}$ by symmetry.

You can prove this result, and work out a formula for the variance, by using two results from pure mathematics.

$$\sum_{r=1}^{n} r = \frac{n(n+1)}{2} \qquad \sum_{r=1}^{n} r^2 = \frac{n}{6}(n+1)(2n+1)$$

These results are described in the C3 module.

To derive the mean, use the general definition $E(X) = \Sigma x p(x)$.

So $E(X) = \sum_{r=1}^{n} r \times \frac{1}{n} = \frac{1}{n} \times \sum_{r=1}^{n} r = \frac{1}{n} \times \frac{n(n+1)}{2} = \frac{(n+1)}{2}$

as expected from the symmetry.

Now to derive the variance, use $E(X^2) = \Sigma x^2 p(x)$.

So

$$E(X^2) = \sum_{r=1}^{n} r^2 \times \frac{1}{n} = \frac{1}{n} \times \sum_{r=1}^{n} r^2 = \frac{1}{n} \times \frac{n}{6}(n+1)(2n+1)$$

$$= \frac{1}{6}(n+1)(2n+1)$$

and

$\text{Var}(x) = E(X^2) - \{E(X)\}^2$

$$= \frac{1}{6}(n+1)(2n+1) - \left(\frac{n+1}{2}\right)^2 = \left(\frac{n+1}{12}\right)\{2(2n+1) - 3(n+1)\}$$

$$= \frac{(n+1)(n-1)}{12}$$

You will not be asked to prove this result, but you do need to be able to use it.

For the discrete uniform distribution with values 1, 2, 3, ..., n

$$E(X) = \mu = \frac{n+1}{2} \qquad \text{Var}(X) = \sigma^2 = \frac{(n+1)(n-1)}{12}$$

EXAMPLE 1

Find the mean and variance of the score on a fair die.

Let X be the score on a fair die.
Then X is a discrete uniform distribution with $n = 6$, so

$$E(X) = \frac{6+1}{2} = 3.5 \quad \text{Var}(X) = \frac{(6+1)(6-1)}{12} = \frac{35}{12}$$

You need to be careful when the values of X do not start at 1.

EXAMPLE 2

The discrete random variable X is a random digit.
Find the mean and variance of X.

X is close to being a discrete uniform distribution with
$n = 10$, but X takes the values 0, 1, …, 9 rather than 1, 2, …, 10.
If Y is a discrete uniform distribution with $n = 10$ then

$$E(Y) = \frac{10+1}{2} = 5.5 \quad \text{Var}(Y) = \frac{(10+1)(10-1)}{12} = \frac{99}{12} = 8.25$$

$$
\begin{aligned}
E(X) &= E(Y-1) & \text{Var}(X) &= \text{Var}(Y-1) \\
&= E(Y) - 1 & &= \text{Var}(Y) \\
&= 4.5 & &= 8.25
\end{aligned}
$$

using the results from Section 7.6.

Exercise 7.7

1 X is a discrete uniform distribution with $n = 4$.
   Find the mean and variance of X.

2 X is a discrete uniform distribution with $n = 8$.
   Find the mean and standard deviation of X.

3 X is a discrete uniform distribution with $n = 6$.
   Find the mean and variance of $2X - 1$.

4 A set of 15 cards has numbers on them.
   A card is chosen at random from the set and the random
   variable X is the number on the card.

   a If the numbers are the integers from 1 to 15 find the mean
      and variance of X.

   b If the numbers are the odd integers from 1 to 29, use your
      answers to part **a** to find the mean and variance of X.

   c If the numbers on the cards are the first 15 prime numbers
      explain why you can not use the answers from part **a** to
      calculate the mean and variance.

S1

1 The random variable, X, has probability function

$$P(X = x) = \frac{k}{x} \quad x = 1, 2, 3, 4$$

a Show that $k = \frac{12}{25}$.

Find

b $P(X < 3)$

c $E(X)$

d $E(5 - 2X)$.

2 The random variable, X, has probability function

$$P(X = x) = \begin{cases} kx & x = 3, 4, 5 \\ k(11 - x) & x = 6, 7, 8 \end{cases}$$

where k is a constant.

a Show that $k = \frac{1}{24}$.

b Find the exact value of $E(X)$.

c Find $Var(X)$.

d Find $Var(732 - 2X)$.

3 A discrete random variable, X, has the probability function shown in the table below.

x	8	10	15
$P(X = x)$	0.4	a	$0.6 - a$

a Given that $E(X) = 10.2$, find a.

b Find $Var(X)$.

c Find $P(X < \mu - \sigma)$.

4 I have an unbiased die. For each of the following state, with a reason, whether or not the random variable is a discrete uniform distribution.

a $X =$ the number of times I roll it until I get a six

b $Y =$ the score showing on the top face when I roll it

c $Z = 7 -$ the score showing on the top face when I roll it

5 A discrete random variable, X, has the cumulative distribution function shown in the table below.

x	6	7	8	9
$F(x)$	0.1	0.3	0.6	1

Find

a $P(X = 8)$

b $E(X)$

c $Var(X)$

d $E(5 - 2X)$

e $Var(5 - 2X)$.

6 a A regular customer in a shop observes that the number of customers, X, in a shop when she enters has the following probability distribution.

No. of customers	0	1	2	3	4
Probability	0.1	0.25	0.3	0.25	0.1

i Find the mean and standard deviation of X.

She also observes that the average waiting time, Y, before being served is as follows:

No. of customers	0	1	2	3	4
Average wait (minutes)	0	2	5	8	11

ii Find the mean waiting time.

b The customer decides that, in future, if there are more than two customers waiting when she arrives she will leave and return another day. On a return visit she will stay whatever the length of the queue.

i What is the probability of her leaving the shop without waiting on her first visit?

ii What is the probability that there are more customers in the shop when she returns for the second visit, on an occasion when she leaves without waiting?

7 The random variable, X, has the probability distribution

x	7	8	9	10	11
$P(X = x)$	0.2	a	0.3	0.1	b

a Given that $E(X) = 9.05$, write down two equations involving a and b.

Find

b the value of a and the value of b

c $Var(X)$

d $Var(5 - 3X)$.

8 The random variable, X, has the probability function

$$P(X = x) = \frac{(11 - 2x)}{25} \quad x = 1, 2, 3, 4, 5$$

a Construct a table giving the probability distribution of X.

Find

b $P(2 < X < 5)$

c $E(X)$

d $Var(X)$

e $Var(2 - 3X)$.

9 The random variable, X, has the cumulative probability distribution:

x	10	12	15	16	18	20	24	25
$P(X \leqslant x)$	0.1	0.3	0.35	0.4	0.6	0.7	0.9	1

a Find the probability distribution of X.

b Find $E(X)$.

c Find $Var(X)$.

d Find $Var(20 - 3X)$.

10 The discrete random variable, X, has probability function

$P(X = x) = kx^2$ for $x = 1, 2, 3, 4$ where k is a positive constant.

a Show that $k = \frac{1}{30}$.

b Find $E(X)$ and show that $E(X^2) = 11.8$.

c Find $Var(X)$.

d Find the mean and variance of $Y = 17 - 4X$.

11 The random variable, X, represents the value of a single-digit random number.

   a Write down the name of the probability distribution of $Y = X + 1$.

   b Calculate the mean and the variance of Y.

A three-digit random number is chosen.

   c Find the probability that it is of the form *ccc*.

12 A discrete random variable, X, has a probability function as shown in the table below, where a and b are constants.

x	1	2	4	8
$P(X = x)$	0.2	0.3	a	b

   a Given that $E(X) = 4.4$, find the value of a and the value of b.

   b Find $P(1 < X < 6)$.

   c Find $E(3X - 4)$.

   d Show that $Var(X) = 9.24$.

   e Evaluate $Var(3X - 4)$.

13 The discrete random variable, X, has probability function

$$P(X = x) = \begin{cases} 0.2 & x = -2 \\ p & x = -1, 0 \\ 0.1 & x = 1, 2 \end{cases}$$

Find

   a p

   b $P(|X| \leqslant 1)$

   c the value of a such that $E(10X + a) = 0$

   d $Var(X)$

   e $Var(10X + a)$.

   f the median m.

The median of X is the value that is halfway. $p(X \leqslant m) = 0.5$

14 The random variable, X, has the discrete uniform distribution

$$P(X = x) = \tfrac{1}{8} \quad x = 1, 2, 3, 4, 5, 6, 7, 8$$

a Write down the value of $E(X)$ and show that $\text{Var}(X) = 5.25$.

Find

b $E(5X - 3)$

c $\text{Var}(7 - 4X)$

15 An independent financial adviser recommends investments in seven traded options.
The number, X, from which the client makes a profit can be modelled by the discrete random variable with probability function

$$P(X = x) = kx \quad x = 0, 1, 2, 3, 4, 5, 6, 7 \quad \text{where } k \text{ is a constant.}$$

a Find the value of k.

b Find $E(X)$ and $\text{Var}(X)$.

The total cost of the investment is £4000 and the return on each successful option is £1500.

c Find the probability that the client makes a loss overall.

d Find the mean and variance of the profit the client makes.

16 A test is taken by a large number of members of the public. There are five questions, and the probability distribution of X, the number of correct answers given, is

x	0	1	2	3	4	5
$P(X = x)$	0.05	0.10	0.25	0.30	0.20	0.10

a Find the mean and standard deviation of X.

A mark, Y, is given where $Y = 10X + 5$.

b Calculate the mean and standard deviation of Y.

17 A supermarket sells top up vouchers valued at £5, £10, £15 or £20. The value, in pounds, of a top up voucher sold may be regarded as a random variable, X, with the following probability distribution

x	5	10	15	20
$P\{X = x\}$	0.20	0.40	0.15	0.25

a Find the mean and standard deviation of X.

b What is the probability that the next top up voucher sold is less than £15?

c As a promotion the marketing department decides the supermarket will offer 10% off the cost of each top up voucher. Write down the mean and standard deviation of the value of vouchers sold in the promotion, assuming customers continue to use the same buying pattern.

The marketing department had originally planned to offer £1 off each voucher.

d i What would have been the mean and standard deviation of the value of vouchers sold, assuming customers continued to use the same buying pattern?

 ii An experienced manager had told the supermarket that they would only sell £5 top up vouchers if they used this strategy. Explain why he thought this would happen.

S1

7

Exit ⟹

Summary

Refer to

- All probabilities must be non-negative and the total probability will always be 1.

  7.1

- A probability function will often have an unknown constant in its expression, which can be found by setting the total probability equal to 1.

  7.2

- $F(x) = P(X \leqslant x)$ is the cumulative distribution function for the random variable X.

  7.3

- The mean or expected value of a probability distribution is defined as $\mu = E(X) = \sum px$

  7.4

- The variance of a probability distribution is defined as $Var(X)$ is $E[\{X - E(X)\}^2] = E(X^2) - \mu^2$ where $E(X^2) = \sum px^2$

  7.5

- The mean and variance of any function of a random variable can be calculated by constructing a new random variable with the values taken by that function, and using the probabilities from the original random variable. If the function is a linear one where $Y = aX + b$, then
  - $E(Y) = a\,E(X) + b$
  - $Var(Y) = a^2\,Var\,(X)$

  7.6

- X follows a discrete uniform distribution if X takes any of the values 1, 2, 3, ... n each with probability $\frac{1}{n}$.

  $E(X) = \frac{n+1}{2}$ and $Var(X) = \frac{(n+1)(n-1)}{12}$

  7.7

Links

The graph shows a simulation of the returns using three investment strategies which *on average* give the same return on the investment, but strategy C is very consistent in the return where B is more variable and A is even more variable.
In the financial services industry, analysis of the random variables in the different strategies allows you to identify differences of this sort, and where millions of dollars are involved in the investment, it can make a big difference.

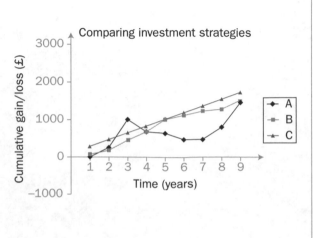

Comparing investment strategies

S1

The Normal distribution

This chapter will show you how to
- identify real-life situations which can be modelled by the Normal distribution
- use probability tables for the standard Normal distribution
- calculate probabilities for a general Normal distribution
- calculate unknown mean and/or standard deviation using given information about probabilities
- use the general Normal distribution in real-life contexts.

Before you start

You should know how to:

1 Solve linear simultaneous equations.

e.g. Solve the simultaneous equations

$$a + 1.25b = 18.25 \qquad (1)$$
$$a - 0.90b = 7.50 \qquad (2)$$

Subtract (2) from (1)

$$2.15b = 10.75$$
$$b = 5$$

Substitute 5 for b in (1)

$$a + 5 \times 1.25 = 18.25$$
$$a = 12$$

2 Substitute values into simple equations.

e.g. Find x when $y = 5$, $a = 3$

and $b = 2$ if $y = \frac{x - a}{h}$.

Substitute for y, a and b:

$$5 = \frac{x - 3}{2}$$
$$10 = x - 3$$
$$x = 13$$

Check in

1 Solve the simultaneous equations

$$36 = \mu - 0.5\sigma$$
$$46 = \mu + 0.5\sigma$$

2 Find p when $a = 16$, $b = 6$, $c = 2$ given

that $p = \frac{a - b}{c}$.

The **Normal distribution** occurs very commonly in real life.

The heights, or weights, of people follow an approximate Normal distribution.

Physicists sometimes refer to this as the **Gaussian distribution**.

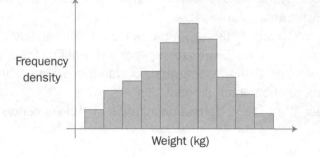

The dimensions of manufactured articles will usually follow a Normal distribution.

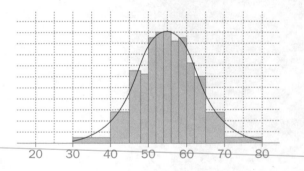

The Normal distribution is a type of **continuous probability distribution**. This means that:

- it relates to a continuous variable (height, weight, etc.)
- it describes the probability of this variable taking a particular range of values.

Generally, the Normal distribution is of the form shown below.

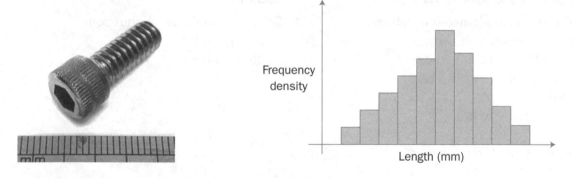

The form of the function is

$$f(x) = \frac{1}{\sigma\sqrt{2\pi}}e^{-\frac{1}{2}\left(\frac{x-\mu}{\sigma}\right)^2}$$

You will not need to learn this formula, or even to use it.

S1

The Normal distribution has the following properties.

- It is symmetrical.
- It is infinite in both directions.
- It has a single peak at the centre.
- It is continuous.
- 95% of values lie within approximately 2 sd of the mean.
- 99% lie within approximately 3 sd of the mean.

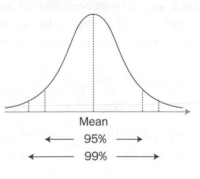

In practice, many real-life variables will not be continuous and will often not be infinite, but the Normal distribution may still be a good approximation.

Also, to enable you to calculate probabilities

The total area under the Normal curve is 1.

The probability that X lies between a and b is the area under the curve between $x = a$ and $x = b$

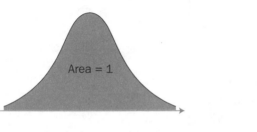

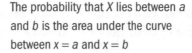

Standardised scores

The Normal distribution allows you to make comparisons between individuals in a particular **normal population**. However, it becomes harder with different normal populations, for example, we use different criteria to judge a 'tall man' than a 'tall woman'.

In 2007 De-Fen Yao was 34 years old and 7ft 8in tall, making her the tallest woman ever.

The tallest recorded man was Robert Pershing Wadlow at 8ft 11^{in} tall – measured in 1940.

One way to compare different normal populations is to standardise the scores – by looking at their distance from the mean, then dividing by the size of one standard deviation.

For example, this process allows comparison of performances of pupils in different subjects, or of athletes under different conditions.

> To find the standardised score, z, from a raw score, x,
>
> use the conversion $z = \dfrac{x - \mu}{\sigma}$
>
> where μ is the mean and σ is the standard deviation of the raw scores.

EXAMPLE 1

In her exams, Alexandra scores 75 in History and 87 in Maths. For the year group as a whole, History has a mean score of 63 on the examination with a standard deviation of 8, while Maths has a mean of 69 with a standard deviation of 15. Compare Alexandra's performance in these two subjects.

For History, $z = \dfrac{75 - 63}{8} = 1.5$

For Maths, $z = \dfrac{87 - 69}{15} = 1.2$

Alexandra's standardised score is higher in History than in Maths, so there is reason to say that her performance is better in History than in Maths.

If you know corresponding points in two distributions, you can work out an unknown mean or standard deviation.

EXAMPLE 2

The mean height of a certain plant, A, is 67 cm, and the heights have a standard deviation of 5 cm. Another plant, B, has a mean height of 63 cm. There is the same proportion of both types of plant taller than 77 cm.
What is the standard deviation of the heights for plant B?

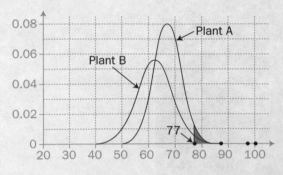

77 cm is 10 cm above the mean for A, or 2 standard deviations. 77 cm is 14 cm above the mean for B and this must also be 2 standard deviations, so $\sigma_B = 7$.

S1

Exercise 8.1

1 $\mu = 56$ $\sigma = 7$

    a Find the standardised score for a raw score of
       i 70   **ii** 52.5   **iii** 66.5   **iv** 56

    b Find the raw score for a standardised score of
       i 1.3   **ii** −2.4   **iii** −0.4   **iv** 2.0

2 $\mu = 87$ $\sigma = 5$

    a Find the standardised score for a raw score of
       i 80   **ii** 59   **iii** 91.3   **iv** 86.7

    b Find the raw score for a standardised score of
       i 2.3   **ii** −2.1   **iii** −0.6   **iv** 1.0

3 $\mu = 3$ $\sigma = 12$

    a Find the standardised score for a raw score of
       i 15   **ii** −3   **iii** −27   **iv** 5.8

    b Find the raw score for a standardised score of
       i 0.7   **ii** −1.3   **iii** −0.2   **iv** 1.8

4 **a** If $\mu = 64$, and 76 has a z score of 2, find σ.

    b If $\sigma = 10$, and 43 has a z score of −1.6, find μ.

5 X has $\mu = 48$ and $\sigma = 10$. Y has mean 53.
   If the same proportions of X and Y are above 68, find the
   standard deviation of Y.

S1

The Normal distribution is written as $X \sim N(\mu, \sigma^2)$.
This means 'X is distributed as a Normal random variable with mean μ and variance σ^2'.

Since all Normal distributions are the same basic shape, you only need to have probabilities for one particular case to allow you to calculate probabilities for all cases.

The standard Normal distribution has mean 0 and variance 1.

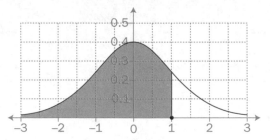

The variable, Z, is often used for the standard Normal distribution.

For the standard Normal distribution, $Z \sim N(0, 1^2)$ or $Z \sim N(0, 1)$.

Be careful to distinguish between variance and standard deviation – if it says $N(83,16)$ you need to use the standard deviation of 4.

By converting values to the standard Normal distribution, you can use probability tables to calculate probabilities for any Normal distribution.

z	$\Phi(z)$	z	$\Phi(z)$	z	$\Phi(z)$	z	$\Phi(z)$	z	$\Phi(z)$
0.00	0.5000	0.50	0.6915	1.00	0.8413	1.50	0.9332	2.00	0.9772
0.01	0.5040	0.51	0.6950	1.01	0.8438	1.51	0.9345	2.02	0.9783
0.02	0.5080	0.52	0.6985	1.02	0.8461	1.52	0.9357	2.04	0.9793
0.03	0.5120	0.53	0.7019	1.03	0.8485	1.53	0.9370	2.06	0.9803
0.04	0.5160	0.54	0.7054	1.04	0.8508	1.54	0.9382	2.08	0.9812

This table is reproduced in full at the end of the book.

The tables give you $P(Z < z)$, where z is a positive value.

To find $P(Z < 1)$:

Locate $z = 1.0$ in the table
Go along to the column 0.00,
and read off the value
$P(Z < 1) = 0.8413$

z	$\Phi(z)$
1.00	0.8413
1.01	0.8438
1.02	0.8461

P(z < 1) = 0.8413

The area beneath the curve represents the probability.

Because of symmetry, the tables only give you half the probabilities – the other half are identical.

So to find $P(Z < -1)$:

$$P(Z < -1) = P(Z > 1)$$
$$= 1 - 0.8413$$
$$= 0.1587$$

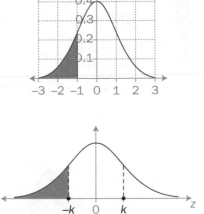

Because you always use the standard Normal distribution to find probabilities it is worth having a special notation for this.
You write $\Phi(z) = P(Z < z)$.
So $\Phi(-k) = 1 - \Phi(k)$ by symmetry.

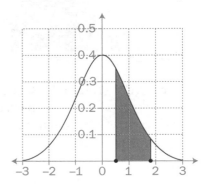

You often need to use more than one value in the tables.

z	Φ(z)
1.78	0.9625
1.79	0.9633
1.80	0.9641
1.81	0.9649
1.82	0.9656
1.83	0.9664

z	Φ(z)
0.50	0.6915
0.51	0.6950
0.52	0.6985

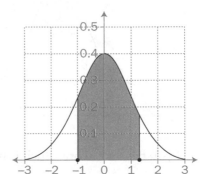

$$P(0.5 < z < 1.8) = \Phi(1.8) - \Phi(0.5) = 0.9641 - 0.6915 = 0.2726$$

You may also need to use a mixture of positive and negative values.

It always helps to draw a clear sketch diagram.

z	Φ(z)
1.00	0.8413
1.01	0.8438
1.02	0.8461

z	Φ(z)
1.28	0.8997
1.29	0.9015
1.30	0.9032
1.31	0.9049
1.32	0.9066

$$P(-1 < z < 1.3) = \Phi(1.3) - \Phi(-1) = 0.9032 - (1 - 0.8413) = 0.7445$$

S1

EXAMPLE 1

If $Z \sim N(0, 1^2)$ find

a P($Z < 1.62$)   **b** P($Z > 0.76$)

c P($Z < -1.32$)   **d** P($-1.2 < Z < 1.7$)

---

a P($Z < 1.62$) = $\Phi(1.62)$ = 0.9474   **b** P($Z > 0.76$) = $1 - \Phi(0.76)$
                                                                            = $1 - 0.7764 = 0.2236$

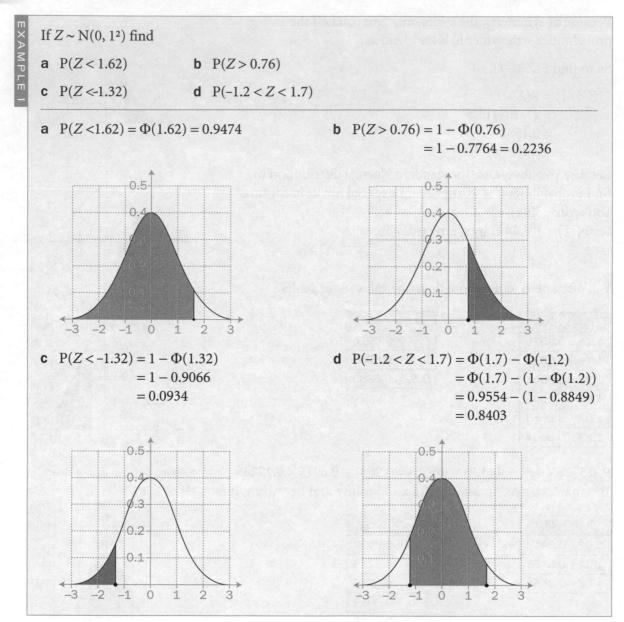

c P($Z < -1.32$) = $1 - \Phi(1.32)$ **d** P($-1.2 < Z < 1.7$) = $\Phi(1.7) - \Phi(-1.2)$
              = $1 - 0.9066$ = $\Phi(1.7) - (1 - \Phi(1.2))$
              = 0.0934 = $0.9554 - (1 - 0.8849)$
                                                               = 0.8403

If you have a z-score which is not listed, such as $z = \dfrac{4}{3}$, you can use linear interpolation to give an approximate answer.

$\dfrac{4}{3} = 1.33333....$

$\Phi(1.33) = 0.9082$ $\Phi(1.34) = 0.9099$

The difference is 0.0017

$\dfrac{1}{3} \times 0.0017 \approx 0.006$ giving $\Phi\left(\dfrac{4}{3}\right) \approx 0.9088$

$$\begin{array}{ccc} & \dfrac{4}{3} & \\ 1.33 & & 1.34 \quad z \\ \hline 0.9082 & ? & 0.9099 \quad \Phi \end{array}$$

You are not required to use interpolation – so using $\Phi(1.33) = 0.9082$ would not be penalised in the examination.

SI

There is a second set of tables provided which give the exact
z-scores for a limited number of tail probabilities. These enable
you to work out the value corresponding to a particular
proportion.

For example, consider the z-value corresponding to the top 5%.

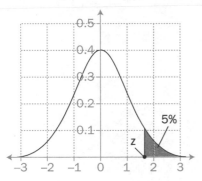

p	z	p	z
0.5000	0.0000	0.0500	1.6449
0.4000	0.2533	0.0250	1.9600
0.3000	0.5244	0.0100	2.3263
0.2000	0.8416	0.0050	2.5758
0.1500	1.0364	0.0010	3.0902
0.1000	1.2816	0.0005	3.2905

This table is reproduced at
the end of this book.

The table shows that $z = 1.6449$.

If you need to find a z-score corresponding to an unlisted
probability, you can use the main table to find an approximation.
For example, consider the z-value corresponding to the top 3.2%

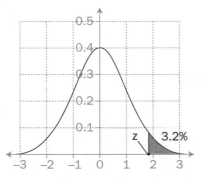

To cut off the top 3.2% you have $\Phi(z) = 0.968$ but there is no
z-value which gives exactly 0.968. Formally, you could write

$$p = 1 - \Phi(z) = 0.032 \Rightarrow \Phi(z) = 0.968$$
$$\Phi(1.85) = 0.9678, \quad \Phi(1.86) = 0.9686$$

and then give $z = 1.85$ as the nearest value or use interpolation to
estimate $z = 1.8525$.

z	$\Phi(z)$
1.83	0.9664
1.84	0.9671
1.85	0.9678
1.86	0.9686
1.87	0.9693

Exercise 8.2

All of these questions relate to the standard Normal distribution,
i.e. $Z \sim N(0, 1^2)$.

1 Find:

   a $\Phi(1.2)$ **b** $\Phi(-0.06)$

   c $\Phi(2.63)$ **d** $\Phi\left(\dfrac{4}{5}\right)$

   e $\Phi(2.5) - \Phi(1.2)$ **f** $\Phi(1.43) - \Phi(-1.03)$

2 Find:

   a $P(Z < 1.08)$ **b** $P(Z > -0.3)$

   c $P(Z < -0.72)$ **d** $P\left(\dfrac{5}{4} < Z < \dfrac{13}{6}\right)$

3 Find the probabilities shown by the shaded areas.

   a

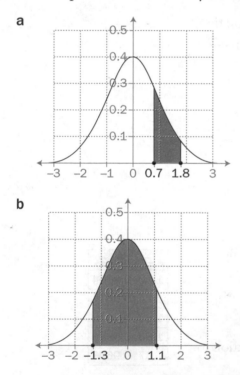

   b

c

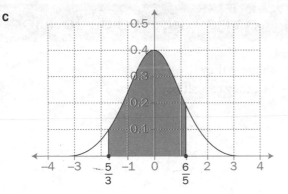

4 Find:

  a $P(|Z| < 1.8)$ **b** $P(|Z| < 0.72)$

  c $P(Z < -2.8 \text{ or } Z > 2.1)$ **d** $P(Z < 1.4 \text{ or } Z > 1.7)$

5 Find the *z*-scores which cut off:

  a the top 10% $P(x > \alpha) = 0.1$ **b** the top 0.5% $P(Z > \alpha) = 0.05$

  c the bottom 2.5% $P(Z < \alpha)$ **d** the bottom 20% $P(-Z < \alpha) = 0.2$
                           $= 0.025$

  e the top 6%   **f** the bottom 1.7%.

S1

All Normal distributions are essentially the same shape – they may have a different centre, or be more peaked – but they can all be standardised to the $N(0, 1)$ distribution.

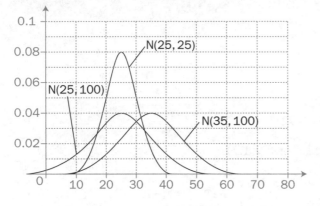

For a distribution $X \sim N(\mu, \sigma^2)$, you can find the probability of X by taking a range of values:

- First calculate z by using $z = \frac{x - \mu}{\sigma}$.
- Then use the table of probabilities to find $\Phi(z)$.
- Deduce the probability you require, referring to a sketch.

S1

EXAMPLE 1

$X \sim N(2, 5^2)$

Find:

a $P(X < 7)$ b $P(X > 7)$ c $P(|X| < 7)$

d $P(|X - 2| < 6)$ e x such that $P(X > x) = 0.05$

- - -

a $P(X < 7)$

$z = \frac{x - \mu}{\sigma} \Rightarrow z = \frac{7 - 2}{5} = 1$

$P(X < 7) = P(Z < 1) = \Phi(1) = 0.8413$

b $P(X > 11)$

$z = \frac{x - \mu}{\sigma} \Rightarrow z = \frac{11 - 2}{5} = 1.8$

$P(X > 11) = P(Z > 1.8) = 1 - \Phi(1.8)$

$\qquad\qquad\qquad = 1 - 0.9641 = 0.0359$

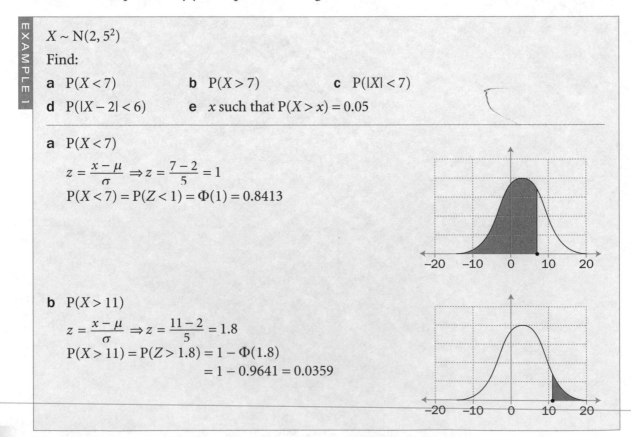

EXAMPLE 1 (CONT.)

c $P(|X| < 3)$

$|x| < 3 \Rightarrow -3 < x < 3$

$z_1 = \dfrac{-3-2}{5} = -1 \qquad z_2 = \dfrac{3-2}{5} = 0.2$

$P(-3 < X < 3) = P(-1 < Z < 0.2) = \Phi(0.2) - \Phi(-1)$

$\qquad\qquad = 0.5793 - (1 - 0.8413) = 0.4206$

d $P(|X - 2| < 6)$

$(|x - 2| < 6) \Rightarrow -4 < x < 8$

$z_1 = \dfrac{-4-2}{5} = -1.2 \qquad z_2 = \dfrac{8-2}{5} = 1.2$

$P(-4 < X < 8) = P(-1.2 < Z < 1.2) = \Phi(1.2) - \Phi(-1.2)$

$\qquad\qquad = 0.8849 - (1 - 0.8849)$

$\qquad\qquad = 0.7698$

e x such that $P(X > x) = 0.05$

$\Phi(z) = 0.95 \Rightarrow z = 1.6449$

$x = \mu + z\sigma$

$\quad = 2 + 1.6449 \times 5$

$\quad = 10.2245$

so $x = 10.2$ (to 3 s.f.)

Use the table of percentage points to find $Z = 1.6449$.

When you are given the probability and have to work out x, use the table of percentage points instead.

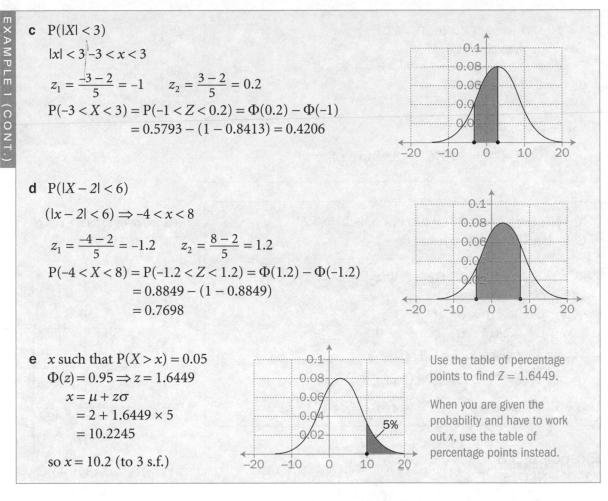

If you are given enough information, you can work out μ or σ or both.

EXAMPLE 2

X is normally distributed such that $X \sim N(\mu, 36)$.
Also, it is known that $P(X > 159.3) = 0.05$.
Calculate the value of μ correct to one decimal place.

$\Phi(z) = 0.95 \Rightarrow z = 1.6449$

$\quad x = \mu + z\sigma$

$159.3 = \mu + 1.6449 \times 6$

$\quad \mu = 159.3 - 1.6449 \times 6$

$\quad\quad = 149.4306$

so $\mu = 149.4$ to 1 d.p.

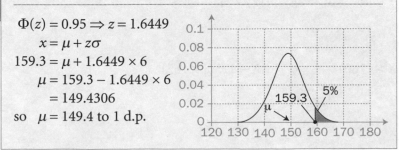

EXAMPLE 3

$X \sim \mathrm{N}(37.1, \sigma^2)$, $\mathrm{P}(X > 51.3) = 0.04$.
Calculate the value of σ.

$\Phi(z) = 0.96 \Rightarrow z = 1.751$
$\quad x = \mu + z\sigma$
$51.3 = 37.1 + 1.751 \times \sigma$

$\sigma = \dfrac{51.3 - 37.1}{1.751} = 8.1096\ldots = 8.11$ to 3 s.f.

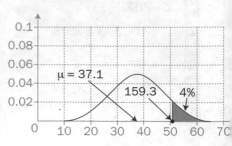

EXAMPLE 4

$X \sim \mathrm{N}(\mu, \sigma^2)$, $\mathrm{P}(X < 37) = 0.1$, $\mathrm{P}(X > 49.3) = 0.2$.
Calculate the values of μ and σ.

$\Phi(z_1) = 0.1 \Rightarrow z_1 = -1.2816 \qquad \Phi(z_2) = 0.8 \Rightarrow z_2 = 0.8416$
Using $x = \mu + z\sigma$ gives
$49.3 = \mu + 0.8416 \times \sigma$
$\quad 37 = \mu - 1.2816 \times \sigma$
Subtracting gives $\quad 12.3 = 2.1232\sigma$

$\sigma = \dfrac{12.3}{2.1232} = 5.793\ldots = 5.79$ (3 s.f.)

$\mu = 37 + 1.2816 \times 5.793\ldots = 44.424\ldots = 44.4$ (3 s.f.)

Here you need to form a pair of simultaneous equations.

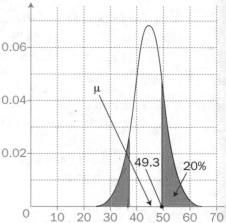

Exercise 8.3

1 $X \sim \mathrm{N}(47, 5^2)$. Find:

  a $\mathrm{P}(X < 56)$ **b** $\mathrm{P}(X > 51)$ **c** $\mathrm{P}(X < 42)$

2 $X \sim \mathrm{N}(32, 16)$. Find:

  a $\mathrm{P}(X < 30)$ **b** $\mathrm{P}(X > 25)$ **c** $\mathrm{P}(X < 30.3)$

3 $X \sim \mathrm{N}(4, 9)$. Find:

  a $\mathrm{P}(X < 10)$ **b** $\mathrm{P}(X > 5)$ **c** $\mathrm{P}(|X| < 3)$

4 $X \sim \mathrm{N}(1750, 165^2)$. Find:

  a $\mathrm{P}(X < 1750)$ **b** $\mathrm{P}(X > 1780)$ **c** $\mathrm{P}(|X - 1750| < 165)$

5 $X \sim \mathrm{N}(0, 20)$. Find:

  a $\mathrm{P}(X < 10)$ **b** $\mathrm{P}(X > 2.5)$ **c** $\mathrm{P}(-3 < X < 5)$

6 $X \sim \mathrm{N}(25, 16)$. Find:

  a x such that $\mathrm{P}(X > x) = 0.05$

  b y such that $\mathrm{P}(X < y) = 0.6$.

7 $X \sim N(83.2, 4.5^2)$

Find:

a x such that $P(X > x) = 0.01$

b y such that $P(X < y) = 0.3$

c z such that $P(X < z) = 0.78$.

8 $X \sim N(0, 15)$

Find:

a x such that $P(|X| < x) = 0.8$

b y such that $P(|X| > y) = 0.6$.

9 $X \sim N(\mu, 5^2)$, $P(X > 23.4) = 0.05$.
Calculate the value of μ.

10 $X \sim N(42, \sigma^2)$, $P(X > 48.3) = 0.01$.
Calculate the value of σ.

11 $X \sim N(\mu, 16)$, $P(X > -3) = 0.98$.
Calculate the value of μ.

12 $X \sim N(186, \sigma^2)$, $P(X < 193) - 0.92$.
Calculate the value of σ.

13 $X \sim N(-32, \sigma^2)$, $P(X < -31.3) = 0.9$.
Calculate the value of σ.

14 $X \sim N(\mu, \sigma^2)$, $P(X < 27) = 0.2$, $P(X > 35) = 0.3$.
Calculate the values of μ and σ.

15 $X \sim N(\mu, \sigma^2)$, $P(X < 78) = 0.6$, $P(X > 89) = 0.2$.
Calculate the values of μ and σ.

16 $X \sim N(\mu, \sigma^2)$, $P(X < 1056) = 0.6$, $P(X > 1132) = 0.2$.
Calculate the values of μ and σ.

17 $X \sim N(\mu, \sigma^2)$, $P(X < 47.3) = 0.5$, $P(X > 52) = 0.2$.
Calculate the values of μ and σ.

The introduction to this chapter refers to the real-life uses of the Normal distribution. This section ties together the techniques covered so far and describes how the distribution is used in practice.

EXAMPLE 1

The lengths of steel girders produced in a factory are normally distributed with a mean length of 12.5 m and a variance of 0.0004 m^2.

Girders need to be between 12.47 and 12.53 metres to be used in construction.

a Find the proportion of girders which cannot be used for construction.

Girders are extremely expensive to produce, and the company is not happy with this level of wastage. A new machine is installed which reduces the variance of the production to 0.0001 m^2.

b Find the proportion of girders produced by the new machine which cannot be used for construction.

In real life, you will often encounter numbers which appear awkward. However, having steel girders almost exactly the same length is necessary for the safety of the building.

a The standard deviation of production is 0.02 m, or 2 cm.
$X \sim N(12.5, 0.02^2)$

$$z_1 = \frac{12.47 - 12.5}{0.02} = -1.5 \quad z_2 = \frac{12.53 - 12.5}{0.02} = 1.5$$

$$\begin{aligned} P(12.47 < X < 12.53) &= P(-1.5 < Z < 1.5) \\ &= \Phi(1.5) - \Phi(-1.5) \\ &= 0.9332 - (1 - 0.9332) \\ &= 0.8664 \end{aligned}$$

13.4% of the girders can not be used.

b The standard deviation of production is now 0.01 m, or 1 cm.
$X \sim N(12.5, 0.01^2)$

$$z_1 = \frac{12.47 - 12.5}{0.01} = -3; \quad z_2 = \frac{12.53 - 12.5}{0.01} = 3$$

$$\begin{aligned} P(12.47 < X < 12.53) &= P(-3 < Z < 3) \\ &= \Phi(3) - \Phi(-3) \\ &= 0.9987 - (1 - 0.9987) \\ &= 0.9974 \end{aligned}$$

0.26% of the girders from the new machine cannot be used.

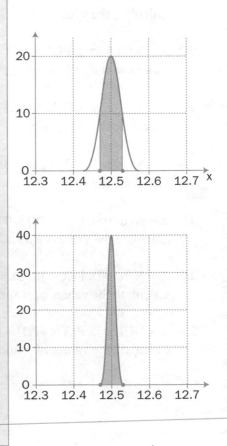

EXAMPLE 2

An examination has a mean mark of 60.7 and a standard deviation of 12.3. You may assume the marks follow a Normal distribution.

a A candidate needs a mark of at least 40 to pass. What percentage of candidates fail?

b The board awards distinctions to the best 10% of the candidates. What is the lowest mark a candidate will need to gain a distinction?

c The list of passing candidates is published the day before the list of distinctions. What is the probability that a candidate who has passed will have a distinction?

---

a $X \sim N(60.7, 12.3^2)$

$$z_1 = \frac{40 - 60.3}{12.3} = -1.6504\ldots$$

$$P(X < 40) = P(-1.6504\ldots < Z)$$
$$= 1 - \Phi(1.65)$$
$$= 1 - 0.9505 = 0.0495$$

So 4.95% of the candidates fail.

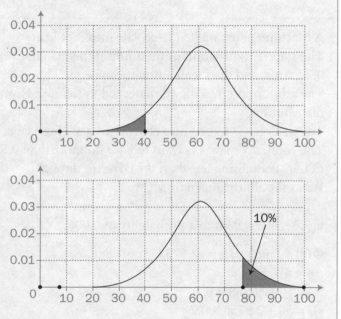

b $X \sim N(60.7, 12.3^2)$

$$\Phi(z) = 0.9 \Rightarrow z = 1.2816$$

$$x = \mu + z\sigma$$
$$= 60.7 + 1.2816 \times 12.3$$
$$= 76.46368$$

So a candidate needs to score at least 77 to gain a distinction.

c This is a conditional probability, but a special sort – the set of candidates who gain a distinction are a subset of the candidates who pass the exam. Therefore P(Distinction ∩ Pass) = P(Distinction), and

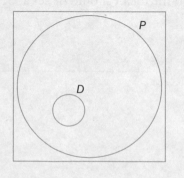

$$P(\text{Distinction}|\text{Pass}) = \frac{P(\text{Distinction})}{P(\text{Pass})}$$

$$= \frac{0.1}{0.9505}$$

$$= 0.1052\ldots$$

$$= 10.5\%$$

S1

Companies producing packaged foods are usually required by law to state the estimated weight of the contents. They need to be able to show that it is rare for their products to contain less than is stated on the packet. However, all production processes are subject to variation.

EXAMPLE 3

A machine pours melted chocolate into moulds. The standard deviation of the amount the machine pours is 0.7 grams, and the mean amount can be set. The amount poured may be assumed to follow a Normal distribution.

The machine is to produce chocolate bars whose label states 50 g. The company's lawyers want to have no more than 0.5% of bars containing less than the advertised weight.

What should the mean be set at?

0.5% is equivalent to a probability of 0.005.

$$\Phi(z) = 0.005 \Rightarrow z = -2.5758$$
$$x = \mu + z\sigma$$
$$50 = \mu - 2.5758 \times 0.7$$
$$\mu = 50 + 2.5758 \times 0.7$$
$$= 51.80306 = 51.8 \text{ (3 s.f.)}$$

So the mean should be set at 51.8 g.

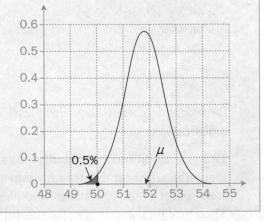

EXAMPLE 4

An airline does a survey of the weights of adult passengers travelling on its flights. It finds that 5% weigh more than 84.3 kg and 2% weigh less than 57.2 kg.
Assuming that the weights of adult passengers on its flights are Normally distributed, find the mean and standard deviation of the weights.

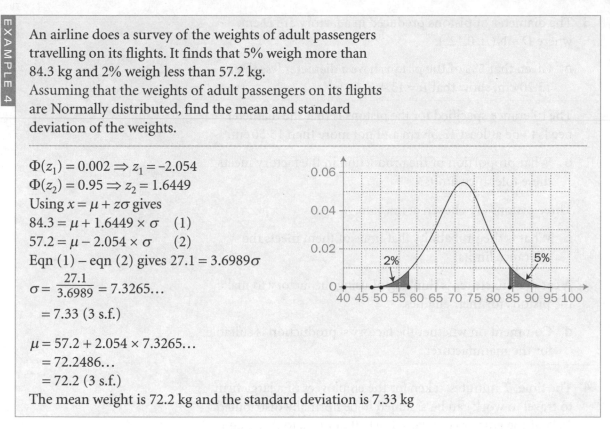

$\Phi(z_1) = 0.002 \Rightarrow z_1 = -2.054$
$\Phi(z_2) = 0.95 \Rightarrow z_2 = 1.6449$
Using $x = \mu + z\sigma$ gives
$84.3 = \mu + 1.6449 \times \sigma$ (1)
$57.2 = \mu - 2.054 \times \sigma$ (2)
Eqn (1) − eqn (2) gives $27.1 = 3.6989\sigma$
$\sigma = \dfrac{27.1}{3.6989} = 7.3265\ldots$
   $= 7.33$ (3 s.f.)

$\mu = 57.2 + 2.054 \times 7.3265\ldots$
   $= 72.2486\ldots$
   $= 72.2$ (3 s.f.)
The mean weight is 72.2 kg and the standard deviation is 7.33 kg

Exercise 8.4

1 IQ is measured on a scale which has a mean of 100 and a standard deviation of 15.
   Find the IQ, X, which is exceeded by only 5% of the population.

2 The lengths, L metres, that Philippe jumps in the triple jump event may be modelled by a normal distribution with mean 15.85 m and variance 0.36 m^2.

   a For a jump chosen at random, find the probability that he jumps at least 16.2 m.

   In a competition, there is a qualifying distance of 16.2 m.

   b Find the probability that he jumps over 16.5 m, given that he makes the qualifying distance with the jump.

3 The diameter of pistons produced in a factory are D cm, where $D \sim \mathrm{N}(\mu, 0.12^2)$.

   a Given that 5% of the pistons have a diameter less than 13.20 cm, show that $\mu = 13.4$.

   The tolerance specified for the pistons is that the diameter needs to be at least 13.35 cm and not more than 13.50 cm.

   b What proportion of the production in the factory meets these tolerance limits?

   Three pistons are chosen at random.

   c What is the probability that none of them meets the tolerance limits?

   A car manufacturer is thinking of using the factory to make the pistons for their engines.

   d Comment on whether the factory's production is suitable for the manufacturer.

4 The time, T minutes, taken for the employees of a large firm to travel to work can be assumed to be normally distributed. 10% of the employees take at least 40 minutes to travel and 0.5% take less than 8 minutes.
   Find the mean and standard deviation of T.

5 The operational life of batteries produced by one manufacturer, A, is normally distributed with mean 43 hours and standard deviation 4 hours.

   a Find the probability that a randomly chosen battery operates for more than 40 hours.

   b What length of operational life is exceeded by 10% of batteries from this manufacturer?

   A rival manufacturer, B, wants to be able to claim that 95% of their batteries last longer than the average produced by manufacturer A. Their process also has a standard deviation of 4 hours.

   c What mean will their manufacturing process need to have to be able to make this claim?

6 A machine fills tins with an amount of liquid which is normally distributed, with mean 330 ml, and standard deviation of 8 ml.

   a Find the probability that a tin contains less than 320 ml.
   b Find the probability that a tin contains between 320 and 345 ml.
   c What volume is exceeded by 5% of the tins?

Another machine also distributes the same liquid with amounts which follow a Normal distribution.

   d If the standard deviation is still 8 ml, what should the mean amount be set to if only 5% of the tins are to be below 320 ml?
   e If the mean is to stay at 330 ml, and only 5% are to be below 320 ml, what would the standard deviation have to be?

7 The weights of eggs from one farm are normally distributed with mean 53 grams and standard deviation 4 grams.

   a What is the probability that an egg from this farm weighs more than 56 grams?

Eggs which weigh less than 48 grams are removed and not sold.

   b What is the probability that an egg sold from this farm weighs more than 56 grams?

8 A company employs a large number of administrative staff. When the company wants to employ new staff, candidates are given a standard task to complete, and the time they take to complete the task is recorded. It is observed that the times taken by candidates are normally distributed with mean 360 seconds and standard deviation 75 seconds.

   a i What proportion of applicants take longer than 450 seconds?
     ii What proportion of the applicants take between 210 seconds and 450 seconds?
     iii What time is exceeded by 5% of the candidates?

Candidates who take longer than 450 seconds are automatically rejected and those who take less than 210 seconds are automatically accepted. The remainder are interviewed.

   b What is the median time taken by those applicants who are interviewed?

9 The volume of oil poured into cans in a production process is known to have a standard deviation of 2.6 ml.
22% of the cans contain less than 660 ml.

   a Calculate the mean volume of oil in the cans.

   b Calculate the proportion of cans which contain less than 650 ml.

1 IQs are measured on a scale with a mean of 100 and standard deviation 15. Assuming that IQs can be modelled by a normal random variable, Find:

  a $P(Y > 130)$

  b $P(73 \leqslant Y \leqslant 91)$

  c the value of k, to 1 decimal place, such that $P(Y \leqslant k) = 0.2$.

2 The performances in the 100 metre sprint of a group of decathletes are modelled by a Normal distribution with mean 11.46 seconds and a standard deviation of 0.32 seconds. The performances of this group of decathletes in the shot put are modelled by a Normal distribution with mean 11.62 m and standard deviation 0.73 m.

  a Find the probability that a randomly chosen athlete:

    i runs the 100 metre sprint faster than 10.8 seconds
    ii puts the shot further than 12.8 m.

  b Assuming that for these decathletes the performances in the two events are independent, find the probability that a randomly chosen athlete runs the 100 metre sprint faster than 10.8 seconds and puts the shot further than 12.8 m.

  c Comment on the assumption that the performances in the two events are independent.

3 Cooking sauces are sold in jars containing a stated weight of 500 g of sauce. The jars are filled by a machine. The actual weight of sauce in each jar is normally distributed with mean 505 g and standard deviation 10 g.

  a i Find the probability that a jar contains less than the stated weight.
    ii In a box of 30 jars, find the expected number of jars containing less than the stated weight.

  The mean weight of sauce is changed so that 1% of the jars contain less than the stated weight. The standard deviation stays the same.

  b Find the new mean weight of sauce.

  [(c) Edexcel Limited 2003]

4 Bottles of spring water have a stated volume of 330 ml. The bottles are filled by a machine, which dispenses volumes which are normally distributed with mean 335 ml and standard deviation 5 ml.

   a Find the probability of a bottle containing less than the stated volume.

The bottles have a capacity of 345 ml. If the machine dispenses more than this, the extra just spills out.

   b Find the probability of a bottle containing less than the stated volume, given that it did not overflow.

A new machine is installed. Only 0.5% of bottles overflow when the mean amount dispensed by the new machine is set to 335 ml.

   c Find the standard deviation of the amount dispensed by the new machine.

5 A random variable, X, has a normal distribution.

   a Describe two features of the distribution of X.

A company produces batteries which have life spans that are normally distributed. Only 2% of the components have a life span less than 230 hours and 5% have a life span greater than 340 hours.

   b Determine the mean and standard deviation of the life spans of the batteries.

The company gives a warranty of 250 hours on the batteries. They make a profit of £7.50 on each battery that they sell. Replacing a battery under warranty costs the company £12.50.

   c Find the average profit they make on the sale of 100 batteries after any replacements under warranty have been made.

6 The duration of the pregnancy of a certain breed of cow is normally distributed with mean μ days and standard deviation σ days. Only 2.5% of all pregnancies are shorter than 235 days and 15% are longer than 286 days.

   a Show that $\mu - 235 = 1.96\sigma$.

   b Obtain a second equation in μ and σ.

   c Find the value of μ and the value of σ.

   d Find the values between which the middle 68.3% of pregnancies lie.   [(c) Edexcel Limited 2002]

S1

7 A certain brand of mushy peas is sold in tins. The mass of mushy peas in a tin is normally distributed with mean 530 g and standard deviation 15 g.

    a   **i**   Find the probability that the mass of mushy peas in a tin is over 555 g.

        ii Find the probability that the mass of mushy peas in a tin is between 522 g and 560 g.

       iii Find the probability that the mass of mushy peas in a tin is between 530 g and 535 g.

    b The kitchen receives some tins from a different manufacturer. The chef knows the two manufacturers use similar machines, so the standard deviation of the masses will be the same, but he does not know the mean mass for these tins. He finds that 10% of the tins from this supplier contain less than 500 grams. Calculate the mean mass of mushy peas in the tins from this manufacturer.   [(c) Edexcel Limited 2001]

8 A health club lets members use, on each visit, its facilities for as long as they wish. The club's records suggest that the length of a visit can be modelled by a normal distribution with mean 90 minutes. Only 20% of members stay for more than 125 minutes.

    a Find the standard deviation of the normal distribution.

    b Find the probability that a visit lasts less than 25 minutes.

The club introduce a closing time of 10.00 p.m.
Tara arrives at the club at 8.00 p.m.

    c Explain whether or not this normal distribution is still a suitable model for the length of her visit.   [(c) Edexcel Limited 2004]

9 The random variable, X, is normally distributed with mean 253 and variance 121.

Find:

    a $P(X < 240)$

    b $P(245 < X < 275)$

It is known that $P(a \leqslant X) = 0.13$. Find the value of a.

10 The length of time it takes an examiner to mark an examination script may be modelled by a Normal distribution with mean 8 minutes and standard deviation 90 seconds. Find the time, t minutes, such that one script in six will take the examiner longer than t minutes to mark.

11 An examination in Statistics consists of a written paper and a project.
Marks for the written paper, E, may be modelled by a normal distribution with mean 62 and standard deviation 9.
Marks for the project, F, may be modelled by a normal distribution with mean 70 and standard deviation 6.

    a Find $P(E > 80)$.   (3)

    b Find p such that $P(F > p) = P(E > 80)$.   (3)

A distinction on the examination requires at least 75 on both the written paper and the project.

    c Find the probability that a candidate gains a distinction, assuming that the performance on the written paper and the project are independent of one another.   (6)

    d Comment on the assumption of independence in part c.   (2)

12 The random variable $X \sim N(\mu, \sigma^2)$.
It is known that

$$P(X \leqslant 69) = 0.0228 \quad \text{and} \quad P(X \geqslant 951) = 0.1056$$

    a i Show that the value of σ is 8.
        ii Find the value of μ.   (8)

    b Find $P(71 \leqslant X \leqslant 81)$.   (3)

13 An electronic component has a useful life span which can be modelled by a Normal distribution with mean 8000 hours and a standard deviation of 400 hours.

    a Calculate the probability that a randomly selected component will last:

        i less than 7700 hours
        ii between 7500 and 8300 hours
        iii at least a year if it is installed on 1 January 2007.   (7)

The components cost the manufacturer £215 to produce.
The manufacturer offers an optional guarantee at an extra cost of £50.
The terms of the guarantee are that the manufacturer will replace the component free if its life is less than 7000 hours, and at a cost of £75 to the customer if its life is between 7000 and 7500 hours.

    b Calculate the expected profit or loss on each guarantee sold by the manufacturer.   (6)

Summary

Refer to

- The Normal distribution is continuous, symmetric, infinite in both directions and has a single peak at the centre.
  - 95% of values lie within approximately 2 sd of the mean
  - 99% lie within approximately 3 sd of the mean

  8.1

- To find the standardised score, z, from a raw score, x, use the conversion $z = \dfrac{x - \mu}{\sigma}$ where μ is the mean and σ is the standard deviation of the raw scores. $x = \mu + z\sigma$ can be used to convert standardised scores back to raw scores.

  8.1

- The probability tables for the N(0,1) distribution give the cumulative probability $\Phi(z)$ for non-negative values of z. The symmetry of the distribution allows the values of $\Phi(z)$ for negative z to be deduced from these. All probabilities can then be worked out using one or two values from the tables.

  8.2

- Calculating probabilities for the $N(\mu, \sigma^2)$ distribution is done by standardising the scores and using the standard Normal distribution tables.
  Calculating an unknown mean and/or standard deviation is done by constructing one or two equations involving the unknowns from the probability information provided and then solving for the unknown(s).

  8.3–8.4

Links

Quality is vitally important in all business practices in today's ultra-competitive global marketplace. Six-sigma (6σ) is a complete quality management approach, developed by the Motorola Corporation) aiming to systematically improve quality through data-driven methods such as statistical process control.

The critical dimensions of many production processes are normally distributed and the term 6σ is derived from the ambition of having less than 1 in a million components not meet the product specification – the proportion lying outside the range $\mu \pm 3\,\sigma$ for the Normal distribution is 1 in a million.

Revision exercise 2

1 Scuba divers carry a tank of compressed air on their backs when they dive. Divers use air faster at greater depths underwater.

    a Explain, with a reason, what sort of correlation would be expected between the maximum depth of a dive, d, and the time, t, a diver spends underwater.

    b Other than the maximum depth, give two quantities you would expect to affect the length of the dive.

2 The summary statistics for a set of observations, (x, y), are:

$$\sum_{i=1}^{n}(x_i - \bar{x})^2 = 533.4 \quad \sum_{i=1}^{n}(y_i - \bar{y})^2 = 371.2 \quad \sum_{i=1}^{n}(x_i - \bar{x})(y_i - \bar{y}) = -315.2$$

Calculate the correlation coefficient for these data. (4)

3 A local historian was studying the number of births in a town and found the following figures relating to the years 1925 to 1934.

Male births, x	223	218	223	223	242	278	299	256	255	292
Female births, y	219	205	209	239	252	256	254	257	259	323

    a Draw a scatter diagram to illustrate this information. (3)

The historian calculated the following summary statistics from the data:

$$S_{xx} = 8276.9 \quad S_{yy} = 10\,230.1 \quad S_{xy} = 7206.3$$

    b Calculate the product moment correlation coefficient.   [(c) Edexcel Limited 1999] (2)

4 When babies are born, the midwife records various physical measurements. The scattergraph shows the head circumference and body length for 16 babies.

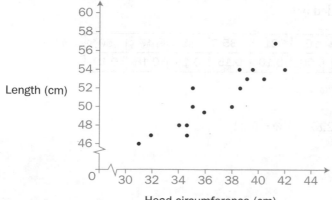

Length (cm)

Head circumference (cm)

Summary statistics for these data are as follows.

$$\sum x = 591 \quad \sum y = 816 \quad \sum x^2 = 22\,001 \quad \sum y^2 = 41\,766 \quad \sum xy = 30\,287$$

a Calculate the correlation coefficient for these data. (4)

Originally, the baby with $x = 32$, $y = 47$ had been entered into the database incorrectly as $x = 47$, $y = 32$, but when the midwife saw the graph, she checked her records and realised there had been a data entry error.

b Calculate what the summary statistics would have been if the error had not been corrected. (4)

c i Show that S_{xy} will be negative for the uncorrected data set. (2)

ii State the sign of the correlation coefficient for the uncorrected data set. (1)

iii Explain why this one data entry error has produced such a large change in the value of the correlation coefficient. (2)

5 An experiment carried out by two students yielded pairs, (x, y), of observations such that

$$\bar{x} = 53.2 \quad \bar{y} = 103.2 \quad S_{xx} = 284.7 \quad S_{xy} = 1204.2$$

a Calculate the equation of the regression line of y on x in the form $y = a + bx$. Give your values of a and b to two decimal places. (3)

b One of the students coded the observations using $u = x - 50$, $v = \dfrac{(y - 100)}{3}$. Find the equation of the line of regression of v on u. (3)

c Estimate the value of y when $x = 50$. (1)

6 In an experiment to discover whether the amount of 'brown fat' in an animal's body affects its tendency to obesity, a biologist took a number of rats from a warm room and put them into a very cold room. The time, in seconds taken for each to start shivering (which is related to the amount of brown fat in the body) was recorded. Nine of these rats were selected, and allowed unlimited access to food for four weeks, after which the increase in their body weight, in kg, was measured. The results are recorded below.

Time to shiver, t	10	10	14	15	21	35	35	47	50
Weight gain, w	0.00	−0.02	0.04	0.06	0.10	0.15	0.18	0.18	0.19

You may assume that

$$\Sigma t = 237 \quad \Sigma t^2 = 8221 \quad \Sigma tw = 32.87 \quad \Sigma w = 0.88$$

a Plot a graph of w against t. (3)

b Find S_{tt} and S_{tw} (4)

c Calculate the equation of the regression line of w on t in the form $w = a + bt$. (3)

d Draw the regression line on your graph. (2)

e Give an interpretation of the values of a and b in this context. (3)

On looking at the graph the biologist conducting the experiment felt that the model was a reasonable one, but that it could be improved.

f Give a reason why the model seems a reasonable one (no calculation is required) and suggest a reason why the biologist thought it could be improved. [(c) Edexcel Limited 1998] (3)

7 The government of a country considered making an investment to decrease the number of members of the population per doctor in order to try and reduce its infant mortality rate. (Infant mortality is measured as the number of infants per 1000 who die before reaching the age of 5.) A study was made of several other similar countries and the variables, x, population per doctor, and y, infant morality, were examined. The data are summarised by the following statistics:

$$\bar{x} = 440.57 \quad \bar{y} = 8.00 \quad S_{xy} = -1598.00 \quad S_{xx} = 174\,567.71$$

a Calculate the equation of the regression line of y on x. (4)

b Given that the country at present has 380 people per doctor, estimate the infant mortality. (2)

c Comment on the coefficient of x in the light of the government's plans. [(c) Edexcel Limited 1999] (3)

8 One measure of personal fitness is the time taken for an individual's pulse rate to return to normal after strenuous exercise: the greater the fitness the slower the time. Reg and Norman have the same normal pulse rates. Following a short programme of strenuous exercise they both recorded their pulse rates, P, at the time, t minutes, after they had stopped exercising. Norman's results are given in the table below.

t	0.5	1.0	1.5	2.0	3.0	4.0	5.0
P	125	113	102	94	81	83	71

a Draw a scatter diagram to represent this information. (4)

The equation of the regression line of P on t for Norman's data is

$$P = 122.3 - 11.0t.$$

b Use the above equation to estimate Norman's pulse rate 2.5 minutes after stopping the exercise programme. (1)

Reg's pulse rate 2.5 minutes after stopping the exercise was 100.

The full data for Reg are summarised by the following statistics:

$$n = 8 \quad \sum t = 19.5 \quad \sum t^2 = 63.75 \quad \sum P = 829 \quad \sum Pt = 1867$$

c   Find the equation of the regression line of P on t for Reg's data. (7)

d   State, giving a reason, which of Reg or Norman you consider
    to be the fitter. [(c) Edexcel Limited 1999] (2)

9 The random variable X has the following probability distribution:

x	1	2	3	4
P(x)	$\frac{1}{4}$	$\frac{1}{2}$	p	$\frac{1}{6}$

where p is a positive constant.

a   Write down the value of p. (1)

b   Find $\mu = \mathrm{E}(X)$. (2)

c   Find $\mathrm{P}(X < \mu)$. [(c) Edexcel Limited specimen] (2)

10 A test is taken by a large number of members of the public.
   There are five questions, and the probability distribution of X,
   the number of correct answers given, is:

x	0	1	2	3	4	5
P(X = x)	0.05	0.10	0.25	0.30	0.20	0.10

a   Find the mean and standard deviation of X. (4)

A mark, Y, is given where $Y = 10X + 5$.

b   Calculate the mean and standard deviation of Y. (4)

11 The discrete random variables, A and B, are independent and
   have the follow probability distributions:

a	1	2	3
P(A = a)	$\frac{1}{4}$	$\frac{1}{2}$	$\frac{1}{4}$

b	1	2
P(B = b)	$\frac{1}{3}$	$\frac{2}{3}$

The random variable, Q, is the product of one observation
from A and one observation from B.

a   Show that $\mathrm{P}(Q = 2) = \frac{1}{3}$. (3)

b   Find the probability distribution for Q. (3)

c   Hence, or otherwise show $\mathrm{E}(Q) = \frac{10}{3}$. (2)

d   Find Var (Q). [(c) Edexcel Limited 1995] (2)

12 An economics student is trying to model the daily movement, X points, in a stock market indicator. The student assumes that the value of X on one day is independent of the value on the next day. A fair die is rolled and if an odd number is uppermost then the indicator is moved down that number of points. If an even number is uppermost then the indicator is moved up that number of points.

  a Write down the distribution of X as specified by the student's model. (2)

  b Find the value of $E(X)$. (2)

  c Show that $\text{Var}(X) = \dfrac{179}{12}$. (3)

  If the indicator moves upwards over a period of time then this is taken as a sign of growth in the economy, if it falls then this is a sign that the economy is in decline.

  d Comment on the state of the economy as suggested by this model. (2)

  Before the stockmarket opened one Monday morning the economic indicator was 3373.

  e Use the student's model to find the probability that the indicator is at least 3400 when the stockmarket closes on the Friday afternoon of the same week. [(c) Edexcel Limited 1998] (3)

13 When a certain type of cell is subjected to radiation, the cell may die, survive as a single cell or divide into two cells with probabilities $\dfrac{1}{2}, \dfrac{1}{3}, \dfrac{1}{6}$ respectively.

  Two cells are independently subjected to radiation. The random variable, X, represents the total number of cells in existence after the experiment.

  a Show that $P(X = 2) = \dfrac{5}{18}$. (3)

  b Find the probability distribution for X. (3)

  c Evaluate $E(X)$. (1)

  d Show that $\text{Var}(X) = \dfrac{10}{9}$. (2)

  Another two cells are submitted to radiation in a similar experiment and the random variable, Y, represents the total number of cells in existence after this experiment. The random variable Z is defined as $Z = X - Y$.

  e Find $E(Z)$ and $\text{Var}(Z)$. [(c) Edexcel Limited 1999] (3)

S1

195

14 The discrete random variable, X, has the probability function shown in the table below.

x	1	2	3	4	5
$P(X = x)$	0.2	0.3	0.3	0.1	0.1

Find:

a $P(2 < X \leqslant 4)$ (2)

b $F(3.7)$ (1)

c $E(X)$ (2)

d $Var(X)$ (4)

e $E(X^2 + 4X - 3)$.   (2)

15 a Write down two reasons for using statistical models.   (2)

    b Give an example of a random variable that could be modelled by

       i a normal distribution

       ii a discrete uniform distribution.   [(c) Edexcel Limited 2006] (2)

16 a Describe one feature of a statistical model which is not necessarily present in a mathematical model.   (2)

    b For each of the following, state whether they could be modelled by a discrete uniform distribution, a normal distribution, or neither.

       i The number of heads obtained when a fair coin is tossed 50 times.

       ii The heights of 12-year-old children in South Africa

       iii The shoe sizes of the children in a Year 4 class.

       iv The total score showing when two fair dice are thrown.   (4)

17 a Describe three features of the normal distribution.   (3)

    b For each of the following, state whether they could be modelled by a discrete uniform distribution, a normal distribution, or neither:

       i the number of sixes showing when you throw three dice

       ii the wingspans of bald eagles

       iii the score showing when you throw a fair die

       iv the number of siblings a child has.   (4)

18 The attendances at home matches of a small town's football team have a mean of 1352 and a standard deviation of 187.

    a Explain why the distribution of X cannot be exactly Normal.   (2)

    b Explain why the Normal distribution may still be a reasonable model for the distribution of X.   (2)

19 The bolts produced by a company have mean length 17.9 mm
   and variance 0.04 mm^2.

   a What proportion of the bolts are less than 17.5 mm? (4)

   Any bolts which are less than 17.5 mm or greater than
   18.4 mm cannot be sold.

   b In a batch of 1000 bolts, how many would you expect to be rejected? (4)

   The mean length can be adjusted by a setting on the machine.

   c At what should the mean be set in order to maximise the
   proportion of bolts satisfying the specifications? (3)

   If a bolt is too short it has to be scrapped, but if it is too long
   it can be filed to make it usable. The mean is set at 17.9 mm.

   d A bolt is chosen at random and found not to be so long that (4)
   it has to be scrapped. What is the probability that it will need
   to be filed before it can be sold?

20 The random variable, X, is normally distributed with mean 157
   and variance 250.

   a Find:
     i $P(X < 140)$ (3)
     ii $P(125 < X < 175)$ (3)

   b It is known that $P(a \leqslant X) = 0.9$.
   Find the value of a. (3)

21 The random variable $X \sim N(\mu, \sigma^2)$.
   It is known that

   $P(X \leqslant -7) = 0.15864$
   and $P(X \geqslant 17) = 0.02275$

   a **i** Show that the value of σ is 8.
      ii Find the value of μ. (8)

   b Find $P(|X| \leqslant 5)$. (3)

22 A machine produces tin plates with a mean length, μ., set by
   the operator and a standard deviation of 1.5 mm.
   On a particular day, 10% of the plates have a length of less
   than 12.2 cm.

   a Calculate, to two decimal places, the value of μ in cm. (4)

   b Calculate the percentage, to one decimal place, of plates which
   will be longer than 12.75 cm on that day. (3)

23 The random variable, X, is normally distributed with mean 560 and standard deviation 20.

   a Find:

     i $P(X < 530)$ (3)

     ii $P(535 < X < 555)$.   (3)

   b It is known that $P(|X - 560| \leqslant a) = 0.6$.
Find the value of a.   (3)

24 The random variable, X, is normally distributed with mean 100 and variance 225.

   a Find:

     i $P(X > 128)$ (3)

     ii $P(76 < X < 124)$.   (4)

   b Find the value of x is such that $P(X < x) = 0.15$.   (3)

The distribution of heights of adults is often modelled by a normal distribution.

   c Give a reason to support the use of normal distribution when modelling height.   (1)

   d Give a reason why such a model may be unrealistic.   [(c) Edexcel Limited 1999] **(1)**

25 A campsite manager recorded the temperature, x, in °C and the number of people, y, swimming in the pool on a sample of days during the summer.

x	24	27	28	29	27	28	24	31	33	27
y	7	15	20	21	23	18	11	30	29	22

     $\sum x = 278$ $\sum x^2 = 7798$ $\sum y = 196$ $\sum y^2 = 4314$ $\sum xy = 5612$

   a Find the equation of the line of regression, y on x.   (6)

   b The temperature is predicted to be 27 °C on a particular day. Estimate the number of people the manager should expect to have swimming in the pool.   (2)

   c The temperature is forecast to reach a maximum of 38 °C on some days in the summer. Explain why the line of regression may not give the manager a reliable indication of how many people will be swimming in the pool on those days.   (2)

Answers

Check in 1

1 **a** **i** 2 **ii** 2 **iii** 5 **iv** 2
   b **i** 6.8 **ii** 6.5 **iii** 2 **iv** 6

Exercise 1.1

1 **a** continuous **b** qualitative **c** discrete
   d qualitative **e** discrete

2

Days late	Frequency
0	19
1	7
2	2
3	1
4	0
5	1
Total	30

3

Depth of water (m.)	Frequency
5.0–5.4	4
5.5–5.9	5
6.0–6.4	4
6.5–6.9	7
7.0–7.4	9
7.5–7.9	1
Total	30

Exercise 1.2

1
```
4 | 8              (1)
5 | 0 0 4 4        (4)
5 | 5 6 7 7 8 8    (6)
6 | 0 0            (2)
6 | 5 5 8          (3)
7 | 1 3 4          (3)
7 | 5 5 6 6 7 8 9  (7)
8 | 0 1 2 3 4      (5)
8 | 5 5 6 7 8 9    (6)
9 | 0 1 3          (3)
```
Key: 6|5 means a time of 65 mins

2
```
29 | 3 7 8    (3)
30 | 1 3      (2)
31 | 1 5 5 9  (4)
32 | 2 4 6    (3)
33 | 7        (1)
34 | 4        (1)
```
Key: 33|7 means head length of 337 mm

3
```
5 | 4                    (1)
5 | 5 5 7 8 9 9 9        (7)
6 | 2 2 2 2 2 3 3 3 4 4  (10)
6 | 5 5 7 7 7 7 8 8 9 9  (10)
7 | 0 1 1 3              (4)
7 | 5 9                  (2)
8 | 0 3                  (2)
```
Key: 5|4 means pulse rate of 54

4
```
0 | 8 9          (2)
1 | 3 4 4 5 7    (5)
2 | 2 7          (2)
3 | 1 2 3        (3)
4 | 1 3          (2)
```
Key: 3|1 means time of 31 minutes

5
```
0 | 0 1 3 3 4 6 7 8 8  (9)
1 | 1 2 2 4 5          (5)
2 | 8                  (1)
3 | 4                  (1)
```
Key: 2|8 means 28 passengers

Exercise 1.3

1 Intervals are 3.45–3.95, 3.95–4.45, 4.45–4.95, 4.95–5.45

2 Intervals are 15–20, 20–25, 25–30, 30–35, 35–40, 40–45, 45–50

3 Intervals are 59.5–69.5, 69.5–79.5, 79.5–89.5, 89.5–99.5, 99.5–109.5, 109.5–119.5

4 Intervals are 79.5–89.5, 89.5–99.5, 99.5–109.5, 109.5–119.5, 119.5–129.5, 129.5–139.5, 139.5–149.5

Exercise 1.4

1 **a** 77 **b** 76.5 **c** 77
2 **a** 4 **b** 4 **c** 4.1 kittens (1 dp)
3 90.8, 109.1 The mean pulse rate rose by 18.3 to 109.1 after warm-up exercise.

4

Length (cm)	Mid-interval, x	Number of plants, f	xf
20–50	35	27	945
50–60	55	18	990
60–65	62.5	16	1000
65–70	67.5	15	1012.5
70–80	75	22	1650
80–90	85	14	1190
90–110	100	14	1400
		$\Sigma f = 126$	$\Sigma xf = 8187.5$

Mean height of plants is $\frac{8187.5}{126} = 64.98$ cm

5 21.09 minutes
6 £16 600

Exercise 1.5

1 76.5, 75, 77
2 $n = 39$ $\frac{1}{4}n = 9.75$ $\frac{1}{2}n = 19.5$ $\frac{3}{4}n = 29.25$,
   so lower quartile is 10th value, median is 20th value, upper quartile is 30th value. $Q_1 = 3$ kittens, $Q_2 = 4$ kittens, $Q_3 = 5$ kittens and IQR $= 5 - 3 = 2$
3 75, 58, 83

Exercise 1.6

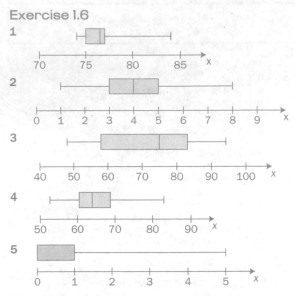

Exercise 2.1

1 e.g. Length, weight, materials to be used, wingspan, means of propulsion etc.

2 e.g. Temperature of water, length of wash cycle, spin speed, number of rinses etc.

3 e.g. Density of traffic, speed of traffic, width of road, whether road is wet, visibility etc.

4 e.g. Number of items, size of items, size of warehouse, fragility of items etc.
e.g. Size of vehicle needed, distance to be travelled, number of drivers/delivery persons needed etc.
e.g. Cost of items, number of days between delivery and payment due

Exercise 2.2

1 A customer's order history can be looked at to create a profile of the categories (s)he appears most interested in, and the most popular titles in those categories across all customers could be recommended. Favourite authors can be treated in a similar way – especially when the author produces a new book.

2 The three scenarios highlight some areas which a negligence award should consider – in particular the loss of earnings. Ali may suffer minimal loss of earnings through the surgeon's negligence, but Wayne's career ending this way will mean a huge loss in earnings – but not known precisely: his future earnings would depend on his own form and fitness, how well his club did, and how long he continued to play at a professional level.

3 e.g. The stength of the radiation it emits and the levels at different locations where you spend a lot of time; generally accepted safety levels and whether the proposal is within those levels; what process the company would have to go through if they want to upgrade the station in the future; what research has been done on the potential health risks for those living near such stations and particularly the time scale of that research.

4 Stage 1 Realisation of a real-world problem.
Stage 6 Statistical techniques are used to test how well the model describes the real-world problem.

Review 2

1 Many real-world problems have uncertainty as an integral part of them – either through imperfect information being inputted or because the same set of circumstances give rise to different outcomes when the 'problem' is observed repeatedly.

2 The sources of uncertainty in a real-world context need to be identified and quantified as far as possible – estimating likelihoods may be done from historical data or by consulting expert opinion in the context. A model is constructed which seeks to replicate the behaviour of the context. The model needs to be tested, sometimes by simulation, before using live runs, to compare its predictions with actual outcomes. The model may need to be adapted in the light of the results of these trials.

3 Simulation is often much cheaper and safer as a first step in testing a model. The computing power available now allows a large number of repetitions of quasi-experiments to be carried out to investigate behaviour of complex situations where a probability analysis of the outcomes would not be possible.

4 A statistical model describes the main features of a real-world situation which has uncertainty as an important feature in it – either because of imperfect information or randomness in the outcomes. The model attempts to quantify the effects of these uncertainties.

5 e.g. Estimating likelihoods (which may be done from historical data or by consulting expert opinion in the context); collecting data to allow comparison of model predictions with actual outcomes; analysis of data – to evaluate the effectiveness of the model in describing the real-world situation.

Check in 3

1 Histogram with bars of heights 0.5, 1.6, 2 and 1.4.

2 **a** 2   **b** 47.3

3 $Q_1 = 2, Q_2 = 3, Q_3 = 4$

Exercise 3.1

1 **a** Type A 44, $47 - 39 = 8$; type B 51, $55 - 49 = 6$
b

(1)	5	2	
(3)	4 2 0	3	
(5)	9 7 7 5 5	3	
(9)	4 4 4 4 2 2 2 2 1	4	4 (1)
(10)	8 7 7 7 7 6 6 6 5 5	4	5 6 7 7 8 8 8 9 9 9 9 (13)
(4)	2 2 1 0	5	0 0 0 0 1 1 1 1 1 2 2 2 3 3 4 (14)
(3)	9 8 7	5	5 5 6 6 7 7 8 9 9 (9)

Key: 6 | 4 | 7 means 46 grams for type A, and 47 grams for type B

c Type B plums are heavier on average than type A, and type B plums are more uniform in weight than type A

d Type B because they are heavier and less variable in size (but type A may be more prolific!)

2   $Q_1 = 72$, $Q_2 = 81$, $Q_3 = 92$; $Q_1 = 87$, $Q_2 = 94$, $Q_3 = 100$
On average, the pulse rates are noticeably higher after exercise than before, and there is less variation in the rates after exercise.

3   $Q_1 = 215$, $Q_2 = 225.5$, $Q_3 = 239$; $Q_1 = 204$, $Q_2 = 215$, $Q_3 = 222$
There is very little difference in the spread of the weights of dunnocks in January and April, but on average the dunnocks are a little heavier in January.

4   $Q_1 = 194$, $Q_2 = 195.5$, $Q_3 = 198$; $Q_1 = 228$, $Q_2 = 232.5$, $Q_3 = 236$
Since the largest male wingspan is less than the smallest female wingspan, it is reasonable to suggest that the males have smaller wingspans than the females. The males have much less variation in their wingspans.

Exercise 3.2

1   a   Males paid more. Half females less than $40 000, only quarter males. Highest earners equal, also lowest almost equal, very few males on lower salaries. Median approximately 5000 more for males.

    b   Starting salaries more equal. Median, lowest and highest equal for genders. As group, females earned more than males on starting. Possibly males promoted more, or males joined staff earlier.

2   More variation in A. More extremes in A. Rainfall. Hours of sun.

3   a

    b   Greater variation in 1990; fewer employees on average in 1995.

Exercise 3.3

1   Frequency densities 0.9, 1.8, 3.2, 3, 2.2, 1.4, 0.7
2   Frequency densities 1.2, 1.4, 2.2, 3.2, 4.4, 3.6, 3, 2.2, 0.6
3   Frequency densities are $115 \div 15 = 7.7$, $46 \div 10 = 4.6$, $36 \div 10 = 3.6$, $22 \div 20 = 1.1$, $14 \div 25 = 0.6$
4   0.7, 16.4, 9, 4.8, 2.6, 0.2
5   a   Relative frequencies 0.06, 0.053, 0.1, 0.07, 0.018
    b   9 pupils

Exercise 3.4

1   a   60.5   b   68   c   71
2   a   28   b   25   c   22
3   a   Cumulative frequencies are 7, 89, 134, 158, 171, 175
      Q_1 is 43.75th value
$$10\,000 + \frac{43.75 - 7}{82} \times 5000 = £12\,240$$
    b D_8 is 140th value $20\,000 + \frac{140 - 134}{24} \times 5000 = £21\,250$

4   a   10.5   b   8
5   a   20   b   39.5   c   29
6   a   27   b   14   c $41 - 9 = 32$

Exercise 3.5

1   a   13.81, 3.86   b   6.38, 1.37
    $\sum xf = 26\,141$ $\sum f = 352$ $\sum x^2 f = 1\,942\,739$

2   $\bar{x} = \frac{26141}{352} = 74.26$ cm

    $\text{var} = \frac{1942739}{352} - \left(\frac{26141}{352}\right)^2 = 3.9728\ldots$

    s.d. $= \sqrt{3.9728\ldots} = 1.99$ cm

3   316.62 mm, 12.17 mm
4   5.114 kg,   0.288 kg
5   2.5,   3.517
6   0.3,   3.43
7   2.76875
8   23.10
9   52, 12

Exercise 3.6

1   $x = 5X + 62.5$ $\bar{x} = 5 \times 3.6 + 62.5 = 80.5$
    variance $= 25 \times 5.3 = 132.5$
2   1137, 322
3   a   55.85   b   10, 12, 11.5, 13, 13, 14, 15.5, 17
    c   13.25   d $\frac{55.85 - 32}{1.8} = 13.25$
4   a   11.167, 2.639   b   0, 1, 2, 3
    c   1.083   0.660
5   a   0, 3, 5, 7, 9, 14   b   4.457   7.477
    c   £16 140   46 731 250
6   a   0, 1, 2, 3   b   1.147   0.8727
    c   18.23 minutes, 21.82 minutes^2

Exercise 3.7

1   a -0.1875 -1.6195 -1.0417 b   Group B
2   a -0.074 -0.186 -0.213 b   Group C
    c -0.339 -0.349 -0.241
    The order is different.
3   a   31.94 years

    b

2	5 6 6 7 7 7 8 9 9	(9)
3	0 2 3 4	(4)
3		(0)
4	2	(1)
4	5	(1)
5	1	(1)

    Key: $2 \,|\, 5$ means 25 years

    c $Q_1 = 27$ $Q_2 = 29$ $Q_3 = 33.5$
    d   IQR $= 33.5 - 27 = 6.5$
      Outliers if $< 27 - 1.5 \times 6.5 = 17.25$ or
      $> 33.5 + 1.5 \times 6.5 = 43.25$,
      so 45 and 51 are outliers.

    e

    f   Positive skew, shape of boxplot or
      $\frac{Q_3 - 2Q_2 + Q_1}{Q_3 - Q_1} = 0.385$ or $Q_3 - Q_2 = 4.5$ and
      $Q_2 - Q_1 = 2$ so $Q_3 - Q_2 > Q_2 - Q_1$

4   a 49.5, $78 - 20 = 58$

  b Fences at −67 and 165, so 182 is outlier

  c Negative skew, shape of boxplot or
      $Q_3 - Q_2 < Q_2 - Q_1$

  d Train at platform when passenger arrives

Exercise 3.8

1 Pulse rates raised after PE. More variation after PE.

2   a 1.1   0.075

  b New diet has greater mean loss and smaller
      variance so volunteers more likely to lose weight
      on new diet

3 Plants in A longer. Longest plant is in A, shortest in B.
    $\frac{1}{4}$ of plants in B shorter than shortest in A. $\frac{1}{4}$ of plants
    in A longer than longest in B.

Review 3

1   a

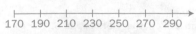

  b positive skew

  c positive skew; longer delays on second day; first
      day half flights less than 4 mins late, second day
      half were over 10 mins late; first day only one
      more than 25 mins late, on second a quarter more
      than 25 mins late, etc.

2

Time (min)	90–139	140–149	150–159	160–169	170–179	180–229
No. of days	8	10	10	4	4	4
Class width	50	10	10	10	10	50
Freq. density	0.16	1	1	0.4	0.4	0.08

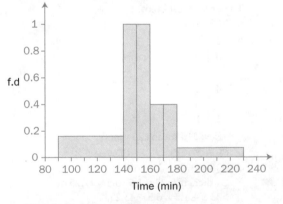

3   a 202   202   233

  b Fences are 146 & 266; 269 is outlier

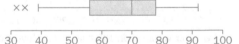

  c Keith's data has positive skew; Asif's almost
      symmetric, slight negative skew

4   a 65.64   31.94   **b** 61 524   **c** 64.69

5

Time (min)	4–7	8	9–10	11	12–16	17–20
Class width	4	1	2	1	5	4
Number of patients	12	20	18	22	15	13
Frequency density	3	20	9	22	3	3.25

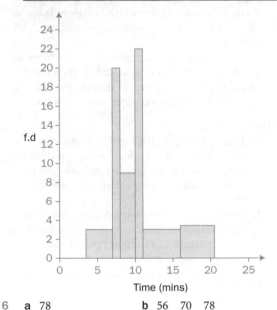

6   a 78   **b** 56   70   78

  c IQR $= 78 - 56 = 22$, $Q_1 - 1.0(Q_3 - Q_1) =$
      $56 - 22 = 34$; $Q_3 + 1.0(Q_3 - Q_1) = 78 + 22 = 100$;
      outliers are 31 and 32

  d mean $= 3363 \div 50 = 67.26$;
      variance $= 238\,305 \div 50 - 67.36^2 = 228.7304$;
      standard deviation $= \sqrt{\text{variance}} = 15.56$

  e $Q_3 - Q_2 = 8$, $Q_2 - Q_1 = 14$, so $Q_3 - Q_2 < Q_2 - Q_1$ and
      mean < median < mode both imply negative skew.

7

```
0 | 0 1 3 3 4    (5)
0 | 6 7 8 8      (4)
1 | 1 2 2 4      (4)
1 | 5            (1)
2 |
2 | 8            (1)
3 | 4            (1)
```

Key: 1|5 means 15 passengers

8   a unequal intervals   **b** 14.5   9.5

  c frequency densities are 6, 5, 2.4, 1.6, 1.1, 0.7

  d 17.90   12.22

  e i   mean decreased, higher values removed

     ii s.d decreased because removed values away
          from mean

9   a 67.6   6.248   **b** $a = 36.36$, $b = 0.6248$

Check in 4

1 **a** $\frac{1}{6}$ **b** $\frac{11}{12}$

2 HH, TT, HT, TH

Exercise 4.1

1 **a** A vowel is not chosen **or** a consonant is chosen.

b A → A I E, B → B, C → C A M B I D G E, D → C E

c $\frac{3}{9}$, $\frac{1}{9}$, $\frac{8}{9}$, $\frac{2}{9}$ **d** A I E   **e** $\frac{3}{9}$

f $\frac{1}{9}$ **g** $\frac{4}{9}$

Exercise 4.2

1

Sum	1	2	3	4	5	6
1	×	3	4	5	6	7
2	3	×	5	6	7	8
3	4	5	×	7	8	9
4	5	6	7	×	9	10
5	6	7	8	9	×	11
6	7	8	9	10	11	×

a $\frac{4}{30}$ **b** $\frac{2}{30}$ **c** 0

2 **a** 1H 1T 2H 2T 3H 3T
4H 4T 5H 5T 6H 6T

b

Sum	1	2	3	4	5	6
1	2	3	4	5	6	7
2	3	4	5	6	7	8

3 **a**

Product	1	2	3	4	5	6
1	1	2	3	4	5	6
2	2	4	6	8	10	12
3	3	6	9	12	15	18
4	4	8	12	16	20	24
5	5	10	15	20	25	30
6	6	12	18	24	30	36

b i $\frac{2}{36}$ **ii** $\frac{2}{36}$

iii $\frac{4}{36}$ **iv** $\frac{2}{36}$

4 **a**

Lower	1	2	3	4	5	6
1	1	1	1	1	1	1
2	1	2	2	2	2	2
3	1	2	3	3	3	3
4	1	2	3	4	4	4
5	1	2	3	4	5	5
6	1	2	3	4	5	6

b i $\frac{7}{36}$ **ii** $\frac{3}{36}$ **iii** $\frac{1}{36}$

5 **a**

Difference	1	2	3	4	5	6
1	0	1	2	3	4	5
2	1	0	1	2	3	4
3	2	1	0	1	2	3
4	3	2	1	0	1	2
5	4	3	2	1	0	1
6	5	4	3	2	1	0

b i $\frac{6}{36}$ **ii** $\frac{2}{36}$ **iii** 0

Exercise 4.3

1 **a**

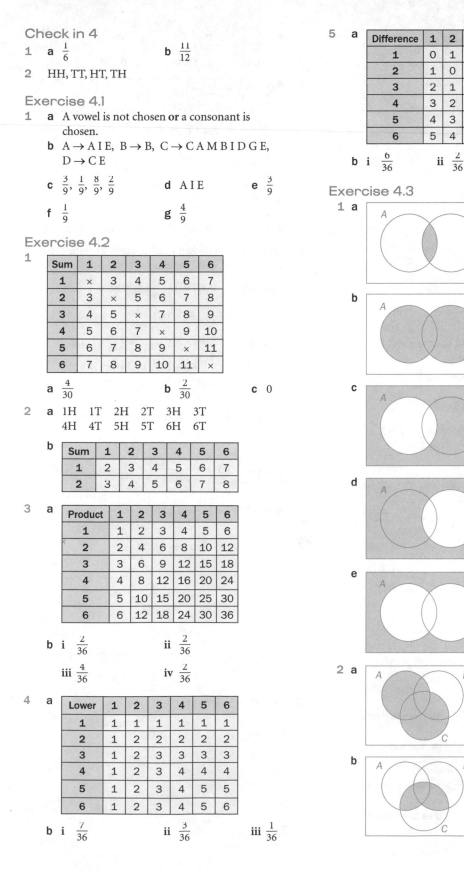

b

c

d

e

2 **a**

b

S1

c

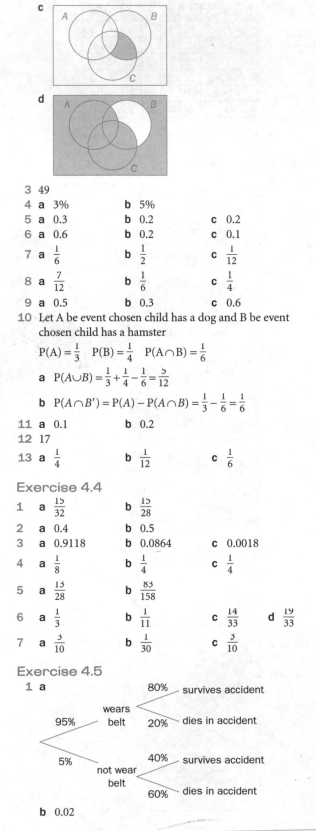

d

3 49

4 a 3%   **b** 5%

5 a 0.3   **b** 0.2   **c** 0.2

6 a 0.6   **b** 0.2   **c** 0.1

7 a $\frac{1}{6}$ **b** $\frac{1}{2}$ **c** $\frac{1}{12}$

8 a $\frac{7}{12}$ **b** $\frac{1}{6}$ **c** $\frac{1}{4}$

9 a 0.5   **b** 0.3   **c** 0.6

10 Let A be event chosen child has a dog and B be event chosen child has a hamster

$P(A) = \frac{1}{3}$ $P(B) = \frac{1}{4}$ $P(A \cap B) = \frac{1}{6}$

a $P(A \cup B) = \frac{1}{3} + \frac{1}{4} - \frac{1}{6} = \frac{5}{12}$

b $P(A \cap B') = P(A) - P(A \cap B) = \frac{1}{3} - \frac{1}{6} = \frac{1}{6}$

11 a 0.1   **b** 0.2

12 17

13 a $\frac{1}{4}$ **b** $\frac{1}{12}$ **c** $\frac{1}{6}$

Exercise 4.4

1 a $\frac{15}{32}$ **b** $\frac{15}{28}$

2 a 0.4   **b** 0.5

3 a 0.9118   **b** 0.0864   **c** 0.0018

4 a $\frac{1}{8}$ **b** $\frac{1}{4}$ **c** $\frac{1}{4}$

5 a $\frac{13}{28}$ **b** $\frac{83}{158}$

6 a $\frac{1}{3}$ **b** $\frac{1}{11}$ **c** $\frac{14}{33}$ **d** $\frac{19}{33}$

7 a $\frac{3}{10}$ **b** $\frac{1}{30}$ **c** $\frac{3}{10}$

Exercise 4.5

1 a

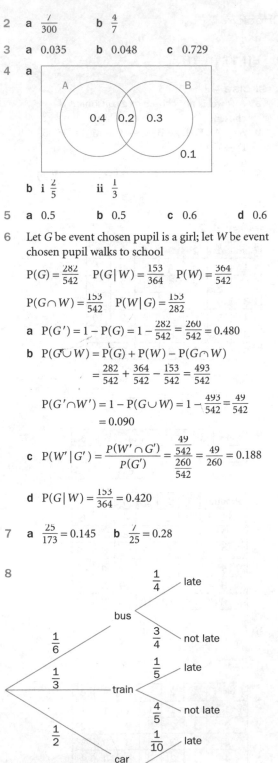

b 0.02

2 a $\frac{7}{300}$ **b** $\frac{4}{7}$

3 a 0.035   **b** 0.048   **c** 0.729

4 a

b i $\frac{2}{5}$ **ii** $\frac{1}{3}$

5 a 0.5   **b** 0.5   **c** 0.6   **d** 0.6

6 Let G be event chosen pupil is a girl; let W be event chosen pupil walks to school

$P(G) = \frac{282}{542}$ $P(G \mid W) = \frac{153}{364}$ $P(W) = \frac{364}{542}$

$P(G \cap W) = \frac{153}{542}$ $P(W \mid G) = \frac{153}{282}$

a $P(G') = 1 - P(G) = 1 - \frac{282}{542} = \frac{260}{542} = 0.480$

b $P(G \cup W) = P(G) + P(W) - P(G \cap W)$

$= \frac{282}{542} + \frac{364}{542} - \frac{153}{542} = \frac{493}{542}$

$P(G' \cap W') = 1 - P(G \cup W) = 1 - \frac{493}{542} = \frac{49}{542}$
$= 0.090$

c $P(W' \mid G') = \frac{P(W' \cap G')}{P(G')} = \frac{\frac{49}{542}}{\frac{260}{542}} = \frac{49}{260} = 0.188$

d $P(G \mid W) = \frac{153}{364} = 0.420$

7 a $\frac{25}{173} = 0.145$ **b** $\frac{7}{25} = 0.28$

8

$\frac{19}{120} = 0.158$

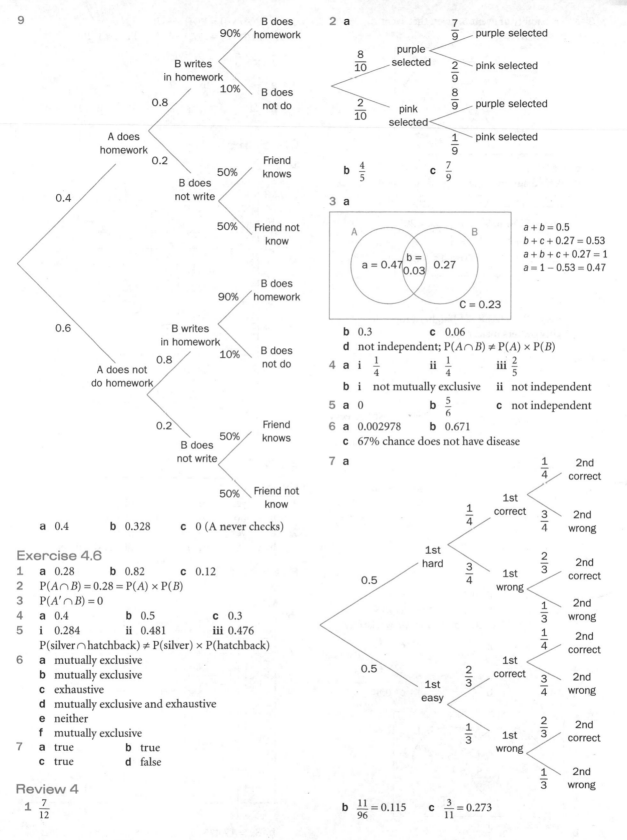

9

B writes in homework 90% B does homework / 10% B does not do

0.8

A does homework

0.2 B does not write 50% Friend knows / 50% Friend not know

0.4

0.6

A does not do homework

B writes in homework 0.8 90% B does homework / 10% B does not do

0.2 B does not write 50% Friend knows / 50% Friend not know

a 0.4   **b** 0.328   **c** 0 (A never checks)

Exercise 4.6
1 **a** 0.28   **b** 0.82   **c** 0.12
2 $P(A \cap B) = 0.28 = P(A) \times P(B)$
3 $P(A' \cap B) = 0$
4 **a** 0.4   **b** 0.5   **c** 0.3
5 **i** 0.284   **ii** 0.481   **iii** 0.476
   $P(silver \cap hatchback) \neq P(silver) \times P(hatchback)$
6 **a** mutually exclusive
   b mutually exclusive
   c exhaustive
   d mutually exclusive and exhaustive
   e neither
   f mutually exclusive
7 **a** true   **b** true
   c true   **d** false

Review 4
1 $\frac{7}{12}$

2 a

$\frac{8}{10}$ purple selected — $\frac{7}{9}$ purple selected / $\frac{2}{9}$ pink selected

$\frac{2}{10}$ pink selected — $\frac{8}{9}$ purple selected / $\frac{1}{9}$ pink selected

b $\frac{4}{5}$ **c** $\frac{7}{9}$

3 a

A: $a = 0.47$ $b = 0.03$ 0.27 : B

C = 0.23

$a + b = 0.5$
$b + c + 0.27 = 0.53$
$a + b + c + 0.27 = 1$
$a = 1 - 0.53 = 0.47$

b 0.3   **c** 0.06
d not independent; $P(A \cap B) \neq P(A) \times P(B)$
4 a i $\frac{1}{4}$ **ii** $\frac{1}{4}$ **iii** $\frac{2}{5}$
   b i not mutually exclusive   **ii** not independent
5 a 0   **b** $\frac{5}{6}$ **c** not independent
6 a 0.002978   **b** 0.671
   c 67% chance does not have disease
7 a

0.5 1st hard — $\frac{1}{4}$ 1st correct — $\frac{1}{4}$ 2nd correct / $\frac{3}{4}$ 2nd wrong
                — $\frac{3}{4}$ 1st wrong — $\frac{2}{3}$ 2nd correct / $\frac{1}{3}$ 2nd wrong

0.5 1st easy — $\frac{2}{3}$ 1st correct — $\frac{1}{4}$ 2nd correct / $\frac{3}{4}$ 2nd wrong
              — $\frac{1}{3}$ 1st wrong — $\frac{2}{3}$ 2nd correct / $\frac{1}{3}$ 2nd wrong

b $\frac{11}{96} = 0.115$ **c** $\frac{3}{11} = 0.273$

8 a probability of event B, given that event A has occurred

b

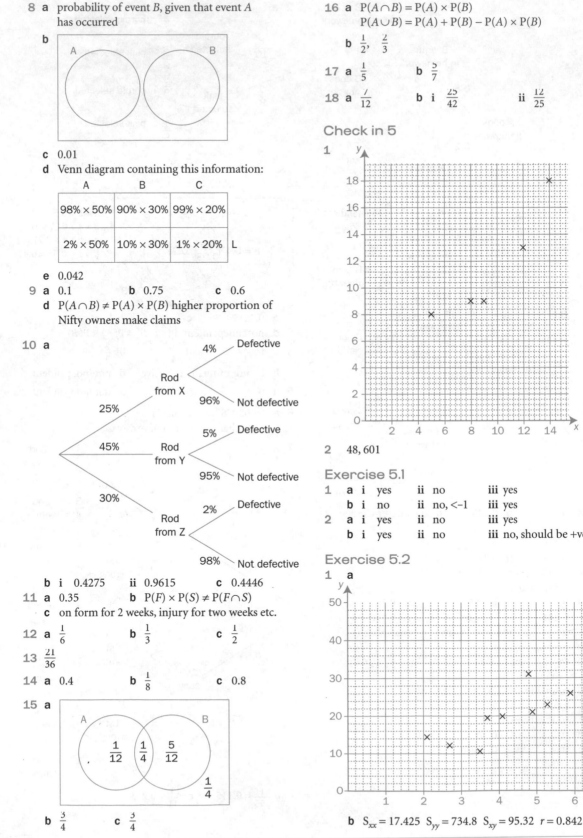

c 0.01

d Venn diagram containing this information:

A	B	C
98% × 50%	90% × 30%	99% × 20%
2% × 50%	10% × 30%	1% × 20% L

e 0.042

9 a 0.1   **b** 0.75   **c** 0.6

d $P(A \cap B) \neq P(A) \times P(B)$ higher proportion of Nifty owners make claims

10 a

b i 0.4275   **ii** 0.9615   **c** 0.4446

11 a 0.35   **b** $P(F) \times P(S) \neq P(F \cap S)$

c on form for 2 weeks, injury for two weeks etc.

12 a $\frac{1}{6}$ **b** $\frac{1}{3}$ **c** $\frac{1}{2}$

13 $\frac{21}{36}$

14 a 0.4   **b** $\frac{1}{8}$ **c** 0.8

15 a

b $\frac{3}{4}$ **c** $\frac{3}{4}$

16 a $P(A \cap B) = P(A) \times P(B)$
$P(A \cup B) = P(A) + P(B) - P(A) \times P(B)$

b $\frac{1}{2}$, $\frac{2}{3}$

17 a $\frac{1}{5}$ **b** $\frac{5}{7}$

18 a $\frac{7}{12}$ **b i** $\frac{25}{42}$ **ii** $\frac{12}{25}$

Check in 5

1

2 48, 601

Exercise 5.1

1 a i yes   **ii** no   **iii** yes
   b i no   **ii** no, <–1   **iii** yes
2 a i yes   **ii** no   **iii** yes
   b i yes   **ii** no   **iii** no, should be +ve

Exercise 5.2

1 a

b $S_{xx} = 17.425$ $S_{yy} = 734.8$ $S_{xy} = 95.32$ $r = 0.842$

2 a

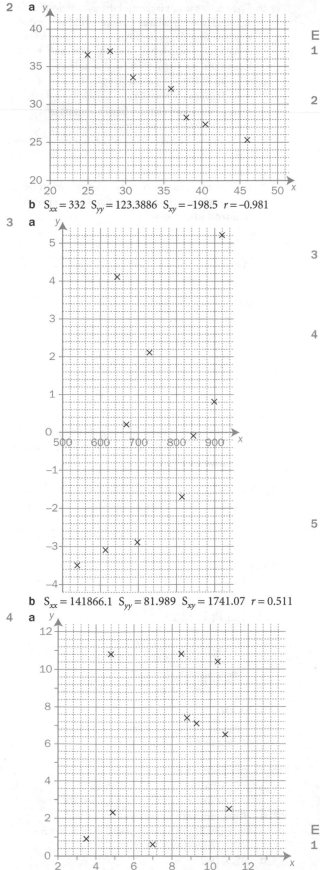

b $S_{xx} = 332$ $S_{yy} = 123.3886$ $S_{xy} = -198.5$ $r = -0.981$

3 a

b $S_{xx} = 141866.1$ $S_{yy} = 81.989$ $S_{xy} = 1741.07$ $r = 0.511$

4 a

b $S_{xx} = 65.856$ $S_{yy} = 152.584$ $S_{xy} = 29.172$ $r = 0.291$

Exercise 5.3

1 a Graph does not support statement.
b $S_{xx} = 479.934...$ $S_{yy} = 84258.9...$
$S_{xy} = -186.857...$ $r = -0.0294$

2 a i $S_{xx} = 0.65184$ $S_{yy} = 250.738...$
$S_{xy} = -3.97356$ $r = -0.311$
ii very weak negative correlation, very little evidence that athletes tend to be good in both disciplines.
b i $S_{xx} = 1.94789$ $S_{yy} = 63.63229$
$S_{xy} = -6.22341$ $r = -0.559$
ii weak negative correlation, some evidence that athletes who are good at long jump do not tend to be good at discus.

3 a $S_{hh} = 280.22...$ $S_{ww} = 124.22...$
$S_{hw} = 130.22...$ $r = 0.698$
b medium positive correlation, some evidence that older men tend to marry older women.

4 a i $S_{dd} = 13.3818...$ $S_{ii} = 1326.72...$
$S_{di} = -88.8636...$ $r = -0.667$
ii medium negative correlation, some evidence that long eruptions are followed by short intervals till next eruption, and vice versa.
b i $S_{ff} = 2289.875$ $S_{nn} = 1481.875$
$S_{fn} = -1626.875$ $r = -0.883$
ii strong negative correlation, evidence that long intervals between eruptions are followed by a short interval, and vice versa.

5 a

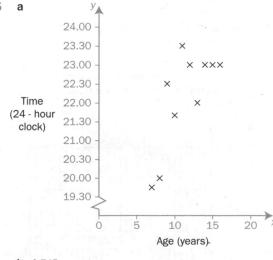

b 0.745
c There is a fairly strong positive correlation.

Exercise 5.4

1 a $S_{xx} = 17.11$ $S_{yy} = 13.857...$ $S_{xy} = 14.5$
b 0.942 **c** 0.942

2 a 62.60, 65.66, 75.92, 73.04, 69.26, 53.24, 65.30, 76.46, 78.80, 94.28, 62.06, 82.22, 83.30, 61.34;
23.228, 93.228, 6.417, 32.913, 28.465, 40.472, 32.087, 53.386, 88.307, 0, 0.354, 0, 62.283, 1.732

b $S_{xx} = 1554.987$ $S_{yy} = 13060.1593$
$S_{xy} = -132.4184$ $r = -0.0294$

c −0.0294

3 a 7, 8, 11, 12, 9, 11, 11, 11, 8, 12; 1.7, 5.9, 4.8, 5.4, 2.8, 5.7, 1.5, 3, 3.8, 4.8

b $S_{xx} = 30$ $S_{yy} = 23.724$ $S_{xy} = 8.1$ $r = 0.304$

c 0.304

4 a 0.9, 1.2, 2.3, 2.1, 2.8, 1.8, 2.4, 0.3, 1.9, 1.3; 5, 6, 6, 5, 6, 3, 7, 4, 1, 8

b $S_{TT} = \sum T^2 - \frac{(\sum T)^2}{10} = 34.18 - \frac{17^2}{10} = 5.28$

$S_{WW} = \sum W^2 - \frac{(\sum W)^2}{10} = 297 - \frac{51^2}{10} = 36.9$

$S_{TW} = \sum TW - \frac{(\sum T)(\sum W)}{10} = 88.5 - \frac{17 \times 51}{10} = 1.8$

$r = \frac{S_{TW}}{\sqrt{S_{TT} \times S_{WW}}} = \frac{1.8}{\sqrt{5.28 \times 36.9}} = 0.129$

c very weak correlation so claim not supported.

d use wider range of times, not just between 9.3 and 11.8.

Review 5

1 *A* −0.79   points in top left and bottom right quadrants
B 0.08   points in all four quadrants
C 0.68   points in top right and bottom left quadrants

2 a $S_{xx} = 155.92\ldots$ $S_{yy} = 214.95\ldots$ $S_{xy} = -157.86\ldots$

b −0.862

c i −0.862

ii fairly strong negative correlation, when *P* has high sales, *Q* has low sales

3 a −0.243

b weak negative correlation, no real evidence of link between front and rear tyres.

c −0.243

4 a

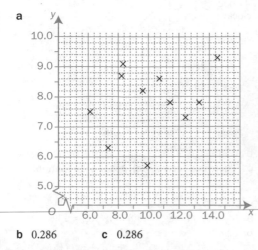

b 0.286   **c** 0.286

d The sample is not large and the correlation is not strong. However a large sample would be needed to argue that the performances were uncorrelated. A data set in which one of the markets suffered a loss would be useful to make this judgement.

5 a $S_{xx} = 3006.222$ $S_{yy} = 904.8889$
$S_{xy} = 1556.444$ $r = 0.944$

b high positive correlation; large amounts of fertilizer associated with high yields.

c changing units codes data, which has no effect on pmcc

6 a 0.870

b fairly strong positive correlation, suggests students who perform well with right hand also perform well with left.

7 a

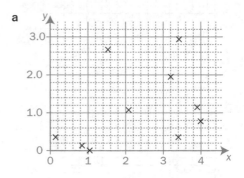

b 0.450

c The correlation is weak but suggests that pupils who trust the police more tend also to trust the press more.

8 a

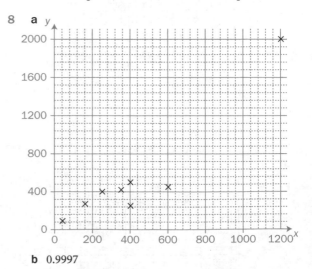

b 0.9997

c Car gives outlier point far away from rest of data so it has a huge effect.

Revision Exercise 1

1 Stage 2. A statistical model is devised
Stage 5. Comparisons are made against the model

2 a Histogram with intervals at:
89.5–139.5, 139.5–149.5, 149.5–159.5, 159.5–169.5, 169.5–189.5

b 147.5, 19.2

c In the 40 working days leading up to the deadline, the senior editor spent much more time in meetings each day than in the other period.

3 a 9.37, 4.40
b the plants in the first garden are smaller on average, with less variation in their height.

4 a $0 + 1 \times 37 + 2 \times 34 + 3 \times 9 = 132$; 254
b mean time = 9.67 minutes, standard deviation = 4.19 minutes.

5 a 45 strawberries
b Boxplot: Min = 38, LQ = 43, Median = 49.5, UQ = 55, Max = 68, Outliers 79, 82
c It is positively skewed (there is a longer tail to the right).

6 a 11, 9, 8, 1
b LQ 6, Median 14, UQ 23
c 21
d Boxplot: Min = 2, LQ = 6, Median = 14, UQ = 23, Max = 54
e strong positive skew; long tail to the right
f Boxplot: Min =15, LQ = 18, Median = 22, UQ = 27, Max = 30
g Trains to Shefton seem to be always late, more so than trains to Darlingborough, but occasionally trains to Darlingborough are delayed by much more than trains to Shefton.

7 a boxplot as described.
b More than a quarter of the pupils in A solved the problems faster than any of the pupils in B, but there was much more variation in the times for class A, who also had the slowest pupils.

8 a £81.45, £104.66
b the data are severely skewed.
c median = £41–50, IQR = (£98.63 – £16.26 =) £82.40
d the median is less than mean, or UQ – median > median – LQ

9 a 7.5
b 1.5
c the mean will be unchanged but the standard deviation will increase since each of these is more than 1 standard deviation away from the (unchanged) mean.

10 a Boxplot: Min = 50, LQ = 61, Median = 73, UQ = 77, Max = 81
b negatively skewed (long tail to the left)
c 68.6, 9.52
d 49.5, 14.85

11 a $\frac{7}{12}$
b i $\frac{25}{42}$ **ii** $\frac{12}{25}$

12 a

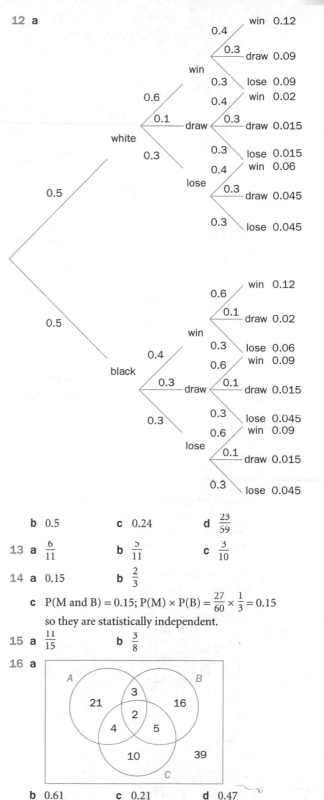

b 0.5 **c** 0.24 **d** $\frac{23}{59}$

13 a $\frac{6}{11}$ **b** $\frac{5}{11}$ **c** $\frac{3}{10}$

14 a 0.15 **b** $\frac{2}{3}$

c P(M and B) = 0.15; P(M) × P(B) = $\frac{27}{60} \times \frac{1}{3} = 0.15$
so they are statistically independent.

15 a $\frac{11}{15}$ **b** $\frac{3}{8}$

16 a

b 0.61 **c** 0.21 **d** 0.47
e $\frac{21}{47} = 0.447$

17 a

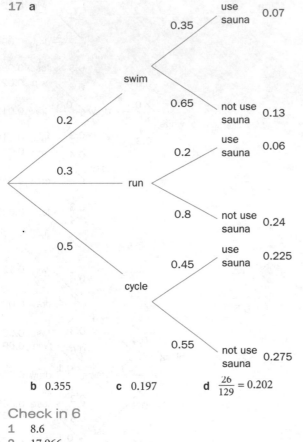

use sauna 0.07
0.35

swim
0.65 not use sauna 0.13

0.2

0.3 run
0.2 use sauna 0.06

0.8 not use sauna 0.24

0.5

cycle
0.45 use sauna 0.225

0.55 not use sauna 0.275

b 0.355 **c** 0.197 **d** $\frac{26}{129} = 0.202$

Check in 6
1 8.6
2 17.966

Exercise 6.1
1 **a** mock = explanatory, final = response
b energy = response, fuel = explanatory
c distance = explanatory, time = response
d cost = response, weight = explanatory
2 **a** reasonable **b** reasonable
c unreasonable; no relationship
d unreasonable; non-linear

Exercise 6.2
2

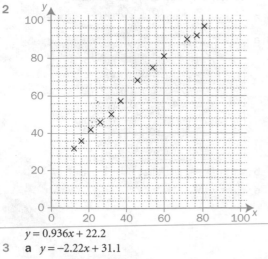

$y = 0.936x + 22.2$
3 **a** $y = -2.22x + 31.1$

b

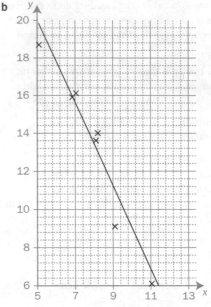

Yes, the line is a good fit to the data.

4 **a** $\bar{p} = \frac{150}{6}$ $\bar{q} = \frac{145}{6}$

$S_{pq} = \sum pq - \frac{(\sum p)(\sum q)}{6}$
$= 3950 - \frac{150 \times 145}{6} = 325$

$S_{pp} = \sum p^2 - \frac{(\sum p)^2}{6} = 4450 - \frac{150^2}{6} = 700$

b $b = \frac{S_{pq}}{S_{pp}} = \frac{325}{700} = 0.464$ $\bar{q} = a + b\bar{p}$

$a = \bar{q} - b\bar{p} = 24.17 - 0.464 \times 25 = 12.56$
equation is $q = 12.56 + 0.464p$

5 **a** $y = -1.42x + 25.60$
6 **a** $\bar{p} = 12.875, \bar{q} = 10.625, S_{pq} = 1068.63, S_{pp} = 1204.88$
b $y = 0.89x - 0.79$
7 $y = 0.815x + 29.22$

Exercise 6.3
1 **a** 123m **b** 97m
c a reliable because 60 within range of data;
b unreliable because 92 outside
2 **a** 32 sec **b** 50 sec
c a reliable – inside range;
b unreliable – outside range
3 **a** 96 min **b** –0.2 min
c impossible answer; unreliable because
outside range
4 **a** $y = 1.12x + 1.54$ **b** 21
c i 34
ii outside range; only 30 symbols in test, so
couldn't remember 34

Exercise 6.4
1 **a** 7, 3, 14, 8, 16, 13, 3, 5; 2, 4, 15, 5, 17, 11, –2, 1
b $Y = 1.28X - 4.42$ **c** $y = 1.28x - 130$
2 **a** $y = 0.143x + 2.15$ **b** $q = 0.715p - 203.8$
3 **a** $S_{CD} = 946.6 - \frac{146.1 \times 38.53}{6} = 8.3945$

$$S_{CC} = 3656.31 - \frac{146.1^2}{6} = 98.775$$

$$b = \frac{S_{CD}}{S_{CC}} = \frac{8.3945}{98.775} = 0.085$$

$$a = \bar{D} - b\bar{C} \quad \bar{D} = \frac{38.53}{6} = 6.422 \quad \bar{C} = \frac{146.1}{6} = 24.35$$

$$a = 6.422 - 0.085 \times 24.35 = 4.352$$

Equation of D on C is $D = 0.085C + 4.352$

$$\frac{d}{3.3} = 0.085 \times \frac{5}{9}(F - 32) + 4.352$$

b $d = 0.085 \times \frac{5}{9} \times 3.3F - 32 \times 0.085 \times \frac{5}{9}$

$\qquad \times 3.3 + 4.352 \times 3.3$

$\qquad d = 0.156F + 9.37$

c $d = 0.156 \times 71 + 9.37 = 20.4$ feet

Review 6

1 a

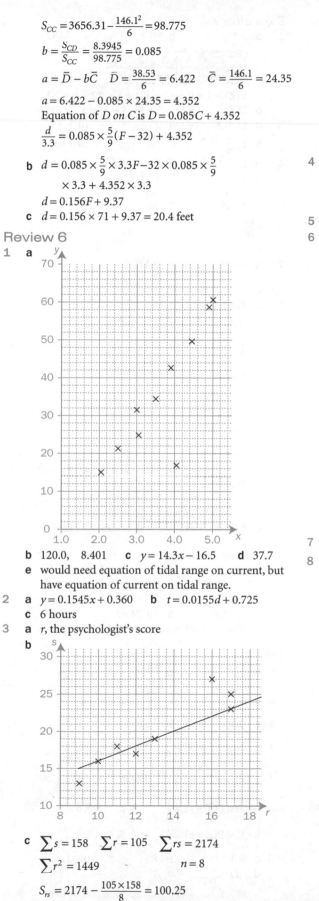

b 120.0, 8.401   **c** $y = 14.3x - 16.5$ **d** 37.7
e would need equation of tidal range on current, but
have equation of current on tidal range.

2 a $y = 0.1545x + 0.360$ **b** $t = 0.0155d + 0.725$
c 6 hours

3 a r, the psychologist's score
b

c $\sum s = 158 \quad \sum r = 105 \quad \sum rs = 2174$

$\quad \sum r^2 = 1449 \qquad\qquad n = 8$

$\quad S_{rs} = 2174 - \frac{105 \times 158}{8} = 100.25$

$$S_{rr} = 1449 - \frac{105^2}{8} = 70.875$$

$$b = \frac{S_{rs}}{S_{rr}} = \frac{100.25}{70.875} = 1.41446$$

$$a = \bar{s} - b\bar{r} = \frac{158}{8} - 1.41446 \times \frac{105}{8} = 1.19$$

$$s = 1.41r + 1.19$$

d see diagram in part **b**
e no, line can only be used to predict stress from
psychologist's score

4 a 357, 1750   **b** $t = 6.83 + 0.204m$ **c** 14.0
d i $m = 120$, outside data set
ii different external temperature, so equation not
valid

5 a $y = 0.441x - 1.16$ **b** $f = 0.441m - 11.4$

6 a

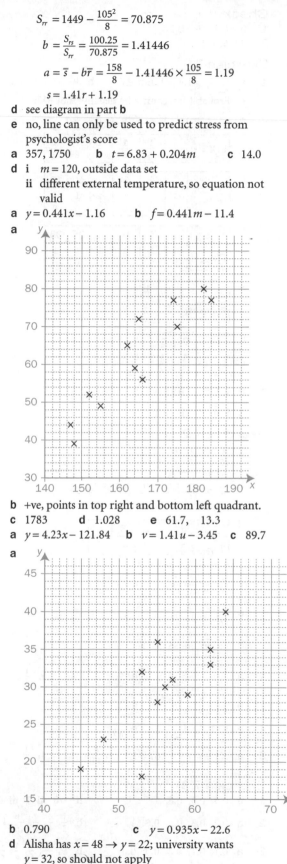

b +ve, points in top right and bottom left quadrant.
c 1783   **d** 1.028   **e** 61.7, 13.3

7 a $y = 4.23x - 121.84$ **b** $v = 1.41u - 3.45$ **c** 89.7

8 a

b 0.790   **c** $y = 0.935x - 22.6$
d Alisha has $x = 48 \rightarrow y = 22$; university wants
$y = 32$, so should not apply
e Karin's GCSE is outside data set; results may not
apply in different school.

Check in 7
1 $a = 0.1$ $b = 0.3$
2 12, 6, 4, 3

Exercise 7.1
1 **a** $\sum p_1 = 1.4 > 1$, not RV   **b** RV   **c** RV
   d Probabilities cannot be negative, not RV

2 **a**

x	1	2	3	4	5	6
$P(X = x)$	$\frac{1}{6}$	$\frac{1}{6}$	$\frac{1}{6}$	$\frac{1}{6}$	$\frac{1}{6}$	$\frac{1}{6}$

   b

y	2	4	6	8	10	12
$P(Y = y)$	$\frac{1}{6}$	$\frac{1}{6}$	$\frac{1}{6}$	$\frac{1}{6}$	$\frac{1}{6}$	$\frac{1}{6}$

   c

z	1	4	9	16	25	36
$P(Z = z)$	$\frac{1}{6}$	$\frac{1}{6}$	$\frac{1}{6}$	$\frac{1}{6}$	$\frac{1}{6}$	$\frac{1}{6}$

   d

w	0	1
$P(W = w)$	$\frac{2}{3}$	$\frac{1}{3}$

3

x	0	1	2
$P(X = x)$	$\frac{1}{4}$	$\frac{1}{2}$	$\frac{1}{4}$

4 **a i** 0.3 **ii** 0.8 **b i** $\frac{1}{6}$ **ii** $\frac{5}{6}$
   c i 0 **ii** 0.1

Exercise 7.2
1 **a** $\frac{1}{15}$ $\frac{2}{15}$ $\frac{3}{15}$ $\frac{4}{15}$ $\frac{5}{15}$
   b $\frac{12}{25}$ $\frac{12}{50}$ $\frac{12}{75}$ $\frac{12}{100}$ *or* $\frac{12}{25}$ $\frac{6}{25}$ $\frac{4}{25}$ $\frac{3}{25}$
   c not RV
   d

r	1	2	3	4
$P(R = r)$	$\frac{30}{77}$	$\frac{20}{77}$	$\frac{15}{77}$	$\frac{12}{77}$

2 **a** $\frac{4}{10}$ $\frac{3}{10}$ $\frac{2}{10}$ $\frac{1}{10}$ **b** $\frac{1}{5}$ $\frac{1}{5}$ $\frac{1}{5}$ $\frac{1}{5}$
   c $\frac{1}{36}$ $\frac{2}{36}$ $\frac{3}{36}$ $\frac{4}{36}$ $\frac{5}{36}$ $\frac{6}{36}$ $\frac{5}{36}$ $\frac{4}{36}$ $\frac{3}{36}$ $\frac{2}{36}$ $\frac{1}{36}$
   d $\frac{1}{36}$ $\frac{3}{36}$ $\frac{5}{36}$ $\frac{7}{36}$ $\frac{9}{36}$ $\frac{11}{36}$

3 **a** $\frac{4}{5}$ **b** $\frac{2}{5}$ **c** $\frac{1}{5}$
   d $\frac{1}{2}$ **e** $\frac{1}{4}$

4 **b i** 0.52   **ii** $\frac{3}{13}$

5 **a** $c(1 + 4 + 9 + 16) = 1$ $c = \frac{1}{30}$ **b** $\frac{1}{6}$

Exercise 7.3
1 **a** 0.3, 0.5, 0.8, 1   **b i** 0 **ii** 1 **iii** 0.5
2 **a** 0.2, 0.3, 0.5, 0.8, 1   **b** $\frac{1}{2}, \frac{3}{4}, \frac{7}{8}, \frac{15}{16}, 1$
3 **a** 0.1, 0.1, 0.2, 0.3, 0.3   **b** $\frac{1}{4}, \frac{1}{12}, \frac{1}{6}, \frac{1}{4}, \frac{1}{4}$

Exercise 7.4
1 **a** 3.2   **b** −0.2   **c** 7.6875
2 **a**

z	1	2	3	4
$P(Z = z)$	0.4	0.3	0.2	0.1

   $E(Z) = 1 \times 0.4 + 2 \times 0.3 + 3 \times 0.2 + 4 \times 0.1 = 2$
   b $E(Y) = 0.2 \times (1 + 2 + 3 + 4 + 5) = 3$
3 **a i** 0.3   **ii** 2.7   **iii** 9.3
   b i $\frac{1}{6}$ **ii** 7.5   **iii** 61.5
4 **a i** 3.1   **ii** 7.3
   b i −0.7   **ii** 3.2
   c i $\frac{51}{16}$ **ii** $\frac{369}{16}$
5 $4\frac{17}{36}$
6 0.4, 0.2
7 **a** 0.175, 0.125   **b** 0.325

Exercise 7.5
1 **a** $E(X) = 7.1$ $Var(X) = 1.29$
   b $E(X) = 0.1$ $Var(X) = 1.29$
2 **a** $\frac{7}{3}, \frac{14}{9}$ **b** $\frac{7}{2}, \frac{35}{12}$ **c** $7, \frac{35}{6}$
3 **a i** 0.3 **ii** 2.9, 1.69   **b i** $\frac{1}{6}$ **ii** $4.5, \frac{11}{12}$
4 2.528, 1.971
5 0.1, 0.5, 2.21
6 0.2, 0.3, 18.01

Exercise 7.6
1 **a** 18.4, 7.6   **b** −13.1, 17.1   **c** 8.7, 1.9
   d 39.9, 93.1
2 **a i** 3.1, 1.29   **ii** 13.5, 32.25
   b i −0.2, 1.36   **ii** 7.4, 5.44
   c i 8.25, 4.8125   **ii** 28.75 43.3125

Exercise 7.7
1 2.5, 1.25
2 4.5, 5.25
3 6, 11.67
4 **a** 8, 18.67   **b** 15, 74.67
   c There is no formula for the prime numbers

Review 7
1 **b** $\frac{18}{25}$ **c** $\frac{48}{25}$ **d** $\frac{29}{25}$
2 **b** $\frac{11}{2}$ **c** $\frac{29}{12}$ **d** $\frac{29}{3}$
3 **a** 0.4   **b** 6.56   **c** 0
4 **a** No; $P(X = 1) = \frac{1}{6}$ $P(X = 2) = \frac{5}{6} \times \frac{1}{6}$

   $P(X = 3) = \left(\frac{5}{6}\right)^2 \times \frac{1}{6}$ $P(X = 4) = \left(\frac{5}{6}\right)^3 \times \frac{1}{6}$ etc

   b Yes; $P(Y = y) = \frac{1}{6}$ $y = 1, 2, 3, 4, 5, 6$
   c Yes; $P(Z = z) = \frac{1}{6}$ $z = 6, 5, 4, 3, 2, 1$
5 **a** 0.3   **b** 8   **c** 1
   d −11   **e** 4

6 a i $E(X) = 0 \times 0.1 + 1 \times 0.25 + 2 \times 0.3 + 3$
$\times 0.25 + 4 \times 0.1 = 2$
$Var(X) = 0 \times 0.1 + 1 \times 0.25 + 4 \times 0.3$
$+ 9 \times 0.25 + 16 \times 0.1 - 2^2 = 1.3$
$sd = \sqrt{Variance} = 1.14$

ii Probabilities are the same as in previous table, eg the waiting time will be 2 minutes if there is 1 customer, $P(Y = 2) = 0.25$

No. of customers, X	0	1	2	3	4
Average wait (mins), Y	0	2	5	8	11
Probability	0.1	0.25	0.3	0.25	0.1

$E(Y) = 0 \times 0.1 + 2 \times 0.25 + 5 \times 0.3 + 8$
$\times 0.25 + 11 \times 0.1 = 5.1$
note: $Y = 3X - 1$ does not hold for $X = 0$

b i Will leave if 3 or 4 waiting, probability = $0.25 + 0.1 = 0.35$

ii Let A be event leaves without waiting on first visit $P(A) = 0.35$
Let B be event more customers on second visit. B will occur if there were 3 waiting on first visit and 4 on second. B is a subset of A, $P(B \cap A) = P(B) = 0.25 \times 0.1 = 0.025$
$P(B|A) = 0.025 \div 0.35 = \frac{1}{14}$

7 a $a + b = 0.4$ $8a + 11b = 3.95$
b 0.15   0.25   **c** 2.0475   **d** 18.4275

8 a

x	1	2	3	4	5
P(X = x)	0.36	0.28	0.2	0.12	0.04

b 0.32   **c** 2.2   **d** 1.36   **e** 12.24

9 a

X	10	12	15	16	18	20	24	25
P(X = x)	0.1	0.2	0.05	0.05	0.2	0.1	0.2	0.1

b 17.85   **c** 26.7275   **d** 240.5475

10 b $\frac{10}{3}$ **c** $\frac{31}{45}$ **d** $\frac{11}{13}, \frac{496}{45} = 11.02$

11 a uniform discrete   **b** 5.5, 8.25   **c** 0.01
12 a 0.1 0.4   **b** 0.4   **c** 9.2   **e** 83.16
13 a 0.3   **b** 0.7   **c** 4   **d** 1.44
   e 144   **f** −1
14 a 4.5   **b** 19.5   **c** 84
15 a $\frac{1}{28}$ **b** 5, 3
   c make loss if 0, 1, 2 are successful. $\frac{3}{28}$
   d $E(1500X - 4000) = 3500$,
   $Var(1500X - 4000) = 6\,750\,000$
16 a 2.8 1.29   **b** 33 12.9
17 a £12.25   £28.69   **b** 0.6
   c £11.03   £23.24
   d i £11.25   £28.69
     ii 2 × £5 vouchers would cost £8 with this discount, 1 × £10 voucher would cost £9. Similarly for other values.

Check in 8
1 $\mu = 41, \sigma = 10$
2 $p = 5$

Exercise 8.1
1 a i 2   **ii** −0.5   **iii** 1.5   **iv** 0
   b i 65.1   **ii** 39.2   **iii** 53.2   **iv** 70
2 a i −1.4   **ii** −5.6   **iii** 0.86   **iv** −0.06
   b i 98.5   **ii** 76.5   **iii** 84   **iv** 92
3 a i 1   **ii** −0.5   **iii** −2.5   **iv** 0.23
   b i 11.4   **ii** −12.6   **iii** 0.6   **iv** 24.6
4 a $\frac{76 - 64}{\sigma} = 2$, $\sigma = 6$ **b** $\frac{43 - \mu}{10} = -1.6$, $\mu = 59$
5 7.5

Exercise 8.2
1 a 0.8849   **b** 0.4761   **c** 0.9957
   d 0.7881   **e** 0.1089   **f** 0.7721
2 a 0.8599   **b** 0.6179   **c** 0.2358
   d 0.0905
3 a 0.2060   **b** 0.7675   **c** 0.8375
4 a 0.9282   **b** 0.5284   **c** 0.0205
   d 0.9638
5 a 1.2816   **b** 2.5758   **c** −1.96
   d −0.8416   **e** 1.555   **f** −2.12

Exercise 8.3
1 a 0.9641   **b** 0.2119   **c** 0.1587
2 a 0.3085   **b** 0.9599   **c** 0.3354
3 a 0.9772   **b** 0.3707   **c** 0.3596
4 a 0.5000   **b** 0.4279   **c** 0.6826
5 a 0.9873   **b** 0.2881   **c** 0.6171
6 a 31.58   **b** 26.01
7 a 93.67   **b** 80.84   **c** 86.67
8 a 4.96   **b** 2.03
9 15.2
10 2.7
11 5.216
12 4.98
13 0.55

14 $P(X < 27) = 0.2 \Rightarrow P\left(Z < \frac{35 - \mu}{\sigma}\right) = 0.2$
$\Rightarrow P\left(Z > \frac{\mu - 27}{\sigma}\right) = 0.2$
$\Rightarrow \frac{\mu - 27}{\sigma} = 0.8416$
$P(X > 35) = 0.3 \Rightarrow P\left(Z > \frac{35 - \mu}{\sigma}\right) = 0.3$
$\Rightarrow \frac{35 - \mu}{\sigma} = 0.5244$
$\mu = \frac{35 \times 0.8416 + 27 \times 0.5244}{0.8416 + 0.5244} = 31.9$
$\sigma = \frac{31.9 - 27}{0.8416} = 5.8$

15 73.3   18.7
16 1023   129
17 47.3   5.58

Exercise 8.4

1 125

2 a 0.280   b 0.498

3 b 0.459   c 0.158

d Less than half their pistons meet the tolerances so it is not suitable with their current production levels.

4 29.4, 8.30

5 a 0.773   b 48.1 hours   c 49.6

6 a 0.1056   b 0.864   c 343 ml

d 333 ml   e 6.1 ml

7 a $X \sim N(53, 4^2)$ $P(X > 56) = P\left(Z > \dfrac{56 - 53}{4}\right)$

$$= P(Z > 0.75)$$
$$= 1 - \phi(0.75)$$
$$= 0.2266$$

b $P(X > 56 | X > 48) = \dfrac{P(X > 56)}{P(X > 48)}$

$$= \dfrac{0.2266}{P\left(Z > \dfrac{48 - 53}{4}\right)}$$

$$= \dfrac{0.2266}{0.8944}$$

$$= 0.253$$

8 a i 0.1151   ii 0.862   iii 483   b 351

9 a 662   b 0

Review 8

1 a 0.0228   b 0.2383   c 87.4

2 a i 0.0196   ii 0.0530   b 0.001

c Very unusual for athlete to be exceptionally good (low probabilities in a and b) in both events, as the disciplines demand very different skills.

3 a i 0.309   ii 9 jars   b 523

4 a 0.1587   b 0.1624   c 3.88

5 a continuous, symmetrical   b 291, 29.7

c £645

6 b $286 - \mu = 1.0364\sigma$ c 268, 17.0

d 251, 285

7 a i 0.0478   ii 0.6803   iii 0.1306

b 519

8 a 41.6   b 0.0591

c More than 20% stay for longer than 120 mins, so having restricted time will affect Tara's visit and this normal distribution will not be a suitable model.

9 a 0.1186   b 0.7437 265.4

10 9.45 minutes

11 a 0.0228   b 82   c 0.015

d Candidate likely to be good at both or bad at both, not independent.

12 a ii 85   b 0.2685

13 a i $P(X < 7700) = \Phi(-0.75) = 0.2266$

ii $P(7500 < X < 8300) = \Phi(0.75) - \Phi(-1.25)$
$$= 0.7734 - 0.1056$$
$$= 0.6678$$

iii $P(X > 8760) = 1 - \Phi(1.9) = 0.0287$

b $P(X < 7000) = \Phi(-2.5) = 0.0062$
$P(7000 < X < 7500) = \Phi(-1.25) - \Phi(-2.5)$
$$= 0.0994$$

Expected cost to manufacturer
$= £215 \times 0.0062 + (£215 - £75) \times 0.0994$
$= £15.25$
Expected profit $= £50 - £15.25 = £34.75$

Revision Exercise 2

1 a Negative correlation, because at greater depths they use the air faster so they can not stay underwater for as long.

b the fitness level of the diver, how long they spend at the maximum depth, how energetic the dive is.

2 −0.708

3 b 0.783

4 a 0.912

b $\sum x = 606, \sum y = 801, \sum x^2 = 23\,186,$
$\sum y^2 = 40\,581, \sum xy = 30\,287$

c i $S_{xy} = -50.875$

ii negative

iii the point (47, 32) is a very long way away from the rest of the data, which are relatively closely bunched together, so the single point has a very large influence.

5 a $y = 4.23x - 121.82$ b $v = 1.41u - 3.45$

c 89.7

6 b 1980, 9.70

c $w = 0.0049t - 0.018$

e a is what the weight gain (loss) would be (on average) in a rat who started to shiver immediately it was transferred; b is the average extra weight gain for each extra second before a rat starts to shiver under these conditions.

f the data are a good fit to the straight line (strong positive correlation), but the data seem to fit a curve better than a straight line.

7 a $y = -0.009154x + 12.0$

b 8.55 deaths per 1000

c the negative coefficient says that as the number of people per doctor increases, the mortality rate decreases which would mean the investment in extra doctors would not be effective – however, the data was collected from a number of different countries, where the infant mortality rate is affected by a lot of other factors as well as the number of people per doctor.

8 b 100   c $P = -9.47t + 126.72$

d Norman is fittest because the gradient of the graph is greatest – hence his pulse returns to normal fastest.

9 a $\dfrac{1}{12}$ b $\dfrac{13}{6}$ c $\dfrac{3}{4}$

10 a 2.8, 1.29   b 33, 12.9

11 a $P(Q = 2) = \frac{1}{2} \times \frac{1}{3} + \frac{1}{4} \times \frac{2}{3} = \frac{1}{3}$

b

q	1	2	3	4	6
$P(Q = q)$	$\frac{1}{12}$	$\frac{1}{3}$	$\frac{1}{12}$	$\frac{1}{3}$	$\frac{1}{6}$

c $E(Q) = 1 \times \frac{1}{12} + 2 \times \frac{1}{3} + 3 \times \frac{1}{12} + 4 \times \frac{1}{3} + 6 \times \frac{1}{6} = \frac{10}{3}$

d $\frac{43}{18} = 2.39$

12 a

X	-1	2	-3	4	-5	6
$P(X = x)$	$\frac{1}{6}$	$\frac{1}{6}$	$\frac{1}{6}$	$\frac{1}{6}$	$\frac{1}{6}$	$\frac{1}{6}$

b 0.5

c $E(X^2) = \frac{1}{6} \times (1 + 4 + 9 + 16 + 25 + 36) = \frac{91}{6}$;

$Var(X) = \frac{91}{6} - \left(\frac{1}{2}\right)^2 = \frac{179}{12}$

d On average the indicator moves up 0.5 points per day, so the model would suggest there is growth in the economy.

e It has to rise at least 27 points in 5 days, so it would have to go up 6 every day – the probability of this is $\left(\frac{1}{6}\right)^5$ in the model, or $\frac{1}{7776} = 0.00013$ – extremely unlikely [not a good model of stock market behaviour]

13 a for X = 2 can have 0, 0 or 1, 1 or 2, 0 for the two cells – with probability $\frac{1}{2} \times \frac{1}{6} + \frac{1}{3} \times \frac{1}{3} + \frac{1}{6} \times \frac{1}{2} = \frac{5}{18}$

b

X	0	1	2	3	4
$P(X = x)$	$\frac{1}{4}$	$\frac{1}{3}$	$\frac{5}{18}$	$\frac{1}{9}$	$\frac{1}{36}$

c $\frac{4}{3}$

d $E(X^2) = \frac{26}{9}$; $\frac{26}{9} - \left(\frac{4}{3}\right)^2 = \frac{10}{9}$

e $0, \frac{20}{9}$

14 a 0.4   **b** 0.8   **c** 2.6

d $8.2 - 2.6^2 = 1.44$ **e** 15.6

15 a capacity to run large scale simulations at low cost; incorporate uncertainty of real-world situations into model – including estimation of unknown inputs; comparison of performance of model predictions with observations;

b i most measurements [height, weight, length etc.] in the natural world can be modelled by the Normal distribution.

ii the score showing when a fair die is thrown.

16 a incorporate uncertainty of real-world situations into model – including estimation of unknown inputs

b i neither – the outcomes are whole numbers from 0 to 50 but they are not equally likely

ii Normal – most measurements [height, weight, length etc.] in the natural world can be modelled by the Normal distribution.

iii neither – there are no fixed limits for the sizes which could be observed

iv neither – the outcomes are not equally likely

17 a it is continuous, symmetric, unimodal, infinite in both directions

b i neither – the outcomes are whole numbers from 0 to 50 but they are not equally likely

ii Normal – most measurements [height, weight, length etc.] in the natural world can be modelled by the Normal distribution.

iii Discrete uniform – takes the values 1, 2, 3, 4, 5 or 6 each with probability $\frac{1}{6}$

iv neither – the outcomes are whole numbers 0, 1, 2, …. but they are not equally likely and there is no identifiable upper limit.

18 a It is not continuous and it can not take negative values.

b There are a very large number of possible values so the approximation of discrete by continuous is not a major problem, and 0 is about 7 standard deviations below the mean so the truncation at zero is negligible.

19 a 0.02275   **b** About 28   **c** 17.95 mm

d $P(X > 18.4 \mid X > 17.5) = \frac{0.00621}{0.97725} = 0.00635$

20 a i 0.859   **ii** 0.851   **b** 177.3

21 a i at 7, $z = -1$, and at 17, $z = 2$, so $3 \times \sigma = 24$, $\sigma = 8$

ii 1

b 0.46483

22 a 12.39   **b** 0.80%

23 a i 0.06681   **ii** 0.296

b 16.8

24 a i 0.031   **ii** 0.890

b 84.5

c most measurements [height, weight, length etc.] in the natural world can be modelled by the Normal distribution.

d A particular subset of a population may have some specific characteristics which mean that the Normal is not a good model e.g. if you were looking at adult basketball players it is likely to be quite skewed where the Normal is a symmetric distribution.

25 a $y = 2.35x - 45.6$

b About 18 people

c 38 lies beyond the range of data therefore can not be used on regression line.

S1

THE NORMAL DISTRIBUTION FUNCTION

The function tabulated below is $\Phi(z)$, defined as $\Phi(z) = \dfrac{1}{\sqrt{2\pi}} \displaystyle\int_{-\infty}^{z} e^{-\frac{1}{2}t^2}$

z	$\Phi(z)$	z	$\Phi(z)$	z	$\Phi(z)$	z	$\Phi(z)$	z	$\Phi(z)$
0.00	0.5000	0.50	0.6915	1.00	0.8413	1.50	0.9332	2.00	0.9772
0.01	0.5040	0.51	0.6950	1.01	0.8438	1.51	0.9345	2.02	0.9783
0.02	0.5080	0.52	0.6985	1.02	0.8461	1.52	0.9357	2.04	0.9793
0.03	0.5120	0.53	0.7019	1.03	0.8485	1.53	0.9370	2.06	0.9803
0.04	0.5160	0.54	0.7054	1.04	0.8508	1.54	0.9382	2.08	0.9812
0.05	0.5199	0.55	0.7088	1.05	0.8531	1.55	0.9394	2.10	0.9821
0.06	0.5239	0.56	0.7123	1.06	0.8554	1.56	0.9406	2.12	0.9830
0.07	0.5279	0.57	0.7157	1.07	0.8577	1.57	0.9418	2.14	0.9838
0.08	0.5319	0.58	0.7190	1.08	0.8599	1.58	0.9429	2.16	0.9846
0.09	0.5359	0.59	0.7224	1.09	0.8621	1.59	0.9441	2.18	0.9854
0.10	0.5398	0.60	0.7257	1.10	0.8643	1.60	0.9452	2.20	0.9861
0.11	0.5438	0.61	0.7291	1.11	0.8665	1.61	0.9463	2.22	0.9868
0.12	0.5478	0.62	0.7324	1.12	0.8686	1.62	0.9474	2.24	0.9875
0.13	0.5517	0.63	0.7357	1.13	0.8708	1.63	0.9484	2.26	0.9881
0.14	0.5557	0.64	0.7389	1.14	0.8729	1.64	0.9495	2.28	0.9887
0.15	0.5596	0.65	0.7422	1.15	0.8749	1.65	0.9505	2.30	0.9893
0.16	0.5636	0.66	0.7454	1.16	0.8770	1.66	0.9515	2.32	0.9898
0.17	0.5675	0.67	0.7486	1.17	0.8790	1.67	0.9525	2.34	0.9904
0.18	0.5714	0.68	0.7517	1.18	0.8810	1.68	0.9535	2.36	0.9909
0.19	0.5753	0.69	0.7549	1.19	0.8830	1.69	0.9545	2.38	0.9913
0.20	0.5793	0.70	0.7580	1.20	0.8849	1.70	0.9554	2.40	0.9918
0.21	0.5832	0.71	0.7611	1.21	0.8869	1.71	0.9564	2.42	0.9922
0.22	0.5871	0.72	0.7642	1.22	0.8888	1.72	0.9573	2.44	0.9927
0.23	0.5910	0.73	0.7673	1.23	0.8907	1.73	0.9582	2.46	0.9931
0.24	0.5948	0.74	0.7704	1.24	0.8925	1.74	0.9591	2.48	0.9934
0.25	0.5987	0.75	0.7734	1.25	0.8944	1.75	0.9599	2.50	0.9938
0.26	0.6026	0.76	0.7764	1.26	0.8962	1.76	0.9608	2.55	0.9946
0.27	0.6064	0.77	0.7794	1.27	0.8980	1.77	0.9616	2.60	0.9953
0.28	0.6103	0.78	0.7823	1.28	0.8997	1.78	0.9625	2.65	0.9960
0.29	0.6141	0.79	0.7852	1.29	0.9015	1.79	0.9633	2.70	0.9965
0.30	0.6179	0.80	0.7881	1.30	0.9032	1.80	0.9641	2.75	0.9970
0.31	0.6217	0.81	0.7910	1.31	0.9049	1.81	0.9649	2.80	0.9974
0.32	0.6255	0.82	0.7939	1.32	0.9066	1.82	0.9656	2.85	0.9978
0.33	0.6293	0.83	0.7967	1.33	0.9082	1.83	0.9664	2.90	0.9981
0.34	0.6331	0.84	0.7995	1.34	0.9099	1.84	0.9671	2.95	0.9984
0.35	0.6368	0.85	0.8023	1.35	0.9115	1.85	0.9678	3.00	0.9987
0.36	0.6406	0.86	0.8051	1.36	0.9131	1.86	0.9686	3.05	0.9989
0.37	0.6443	0.87	0.8078	1.37	0.9147	1.87	0.9693	3.10	0.9990
0.38	0.6480	0.88	0.8106	1.38	0.9162	1.88	0.9699	3.15	0.9992

THE NORMAL DISTRIBUTION FUNCTION (continued)

z	$\Phi(z)$	z	$\Phi(z)$	z	$\Phi(z)$	z	$\Phi(z)$	z	$\Phi(z)$
0.39	0.6517	0.89	0.8133	1.39	0.9177	1.89	0.9706	3.20	0.9993
0.40	0.6554	0.90	0.8159	1.40	0.9192	1.90	0.9713	3.25	0.9994
0.41	0.6591	0.91	0.8186	1.41	0.9207	1.91	0.9719	3.30	0.9995
0.42	0.6628	0.92	0.8212	1.42	0.9222	1.92	0.9726	3.35	0.9996
0.43	0.6664	0.93	0.8238	1.43	0.9236	1.93	0.9732	3.40	0.9997
0.44	0.6700	0.94	0.8264	1.44	0.9251	1.94	0.9738	3.50	0.9998
0.45	0.6736	0.95	0.8289	1.45	0.9265	1.95	0.9744	3.60	0.9998
0.46	0.6772	0.96	0.8315	1.46	0.9279	1.96	0.9750	3.70	0.9999
0.47	0.6808	0.97	0.8340	1.47	0.9292	1.97	0.9756	3.80	0.9999
0.48	0.6844	0.98	0.8365	1.48	0.9306	1.98	0.9761	3.90	1.0000
0.49	0.6879	0.99	0.8389	1.49	0.9319	1.99	0.9767	4.00	1.0000
0.50	0.6915	1.00	0.8413	1.50	0.9332	2.00	0.9772		

PERCENTAGE POINTS OF THE NORMAL DISTRIBUTION

The values z in the table are those which a random variable $Z \sim N(0, 1)$ exceeds with probability p; that is, $P(Z > z) = 1 - \Phi(z) = p$.

p	z	p	z
0.5000	0.0000	0.0500	1.6449
0.4000	0.2533	0.0250	1.9600
0.3000	0.5244	0.0100	2.3263
0.2000	0.8416	0.0050	2.5758
0.1500	1.0364	0.0010	3.0902
0.1000	1.2816	0.0005	3.2905

Formulae

Students will be given the following formulae in the exam formula booklet (entitled Mathematical Formulae including Statistical Formulae and Tables):

Probability

$$P(A \cup B) = P(A) + P(B) - P(A \cap B)$$
$$P(A \cap B) = P(A)P(B \mid A)$$

$$P(A \mid B) = \frac{P(B \mid A)P(A)}{P(B \mid A)P(A) + P(B \mid A')P(A')}$$

Discrete distributions

For a discrete random variable X taking values x_i with probabilities $P(X = x_i)$

Expectation (mean): $E(X) = \mu = \sum x_i P(X = x_i)$

Variance: $\mathrm{Var}(X) = \sigma^2 = \sum (x_i - \mu)^2 P(X = x_i) = \sum x_i^2 P(X = x_i) - \mu^2$

For a function $g(X)$: $E(g(X)) = \sum g(x_i) P(X = x_i)$

Continuous distributions

Standard continuous distribution:

Distribution of X	P.D.F.	Mean	Variance
Normal $N(\mu, \sigma^2)$	$\dfrac{1}{\sigma\sqrt{2\pi}} e^{-\frac{1}{2}\left(\frac{x-\mu}{\sigma}\right)^2}$	μ	σ^2

Correlation and regression

For a set of n pairs of values (x_i, y_i)

$$S_{xx} = \sum (x_i - \bar{x})^2 = \sum x_i^2 - \frac{(\sum x_i)^2}{n}$$

$$S_{yy} = \sum (y_i - \bar{y}^2) = \sum y_i^2 - \frac{(\sum y_i)^2}{n}$$

$$S_{xy} = \sum (x_i - \bar{x})(y_i - \bar{y}) = \sum x_i y_i - \frac{(\sum x_i)(\sum y_i)}{n}$$

The product moment correlation coefficient is

$$r = \frac{S_{xy}}{\sqrt{S_{xx} S_{yy}}} = \frac{\sum (x_i - \bar{x})(y_i - \bar{y})}{\sqrt{\{\sum (x_i - \bar{x})^2\}\{\sum y_i - \bar{y})^2\}}} = \frac{\dfrac{\sum x_i y_i - (\sum x_i)(\sum y_i)}{n}}{\sqrt{\left(\sum x_i^2 - \dfrac{(\sum x_i)^2}{n}\right)\left(\sum y_i^2 - \dfrac{(\sum y_i)^2}{n}\right)}}$$

The regression coefficient of y on x is $b = \dfrac{S_{xy}}{S_{xx}} = \dfrac{\sum (x_i - \bar{x})(y_i - \bar{y})}{\sum (x_i - \bar{x})^2}$

Least squares regression line of y on x is $y = a + bx$ where $a = \bar{y} - b\bar{x}$

In addition to the above, students will be expected to remember the following formulae, which will **not** be given in the exam formula booklet:

Mean $= \bar{x} = \dfrac{\Sigma x}{n}$ or $\dfrac{\Sigma fx}{\Sigma f}$

Standard deviation $= \sqrt{(\text{Variance})}$

Interquartile range $= \text{IQR} = Q_3 - Q_1$

$P(A') = 1 - P(A)$

For independent events A and B

$P(B|A) = P(B), P(A|B) = P(A)$

$P(A \cap B) = P(A)\,P(B)$

$E(aX + b) = aE(X) + b$

$\text{Var}(aX + b) = a^2\,\text{Var}(X)$

Cumulative distribution function for a discrete random variable:

$F(x_0) = P(X \leqslant x_0) = \displaystyle\sum_{x \leqslant x_n} p(x)$

Standardised Normal Random Variable $Z = \dfrac{X - \mu}{\sigma}$
where $X \sim N(\mu, \sigma^2)$

Glossary

Boxplot A graphical representation of a distribution using only the minimum, maximum, median and the lower and upper quartiles of the data.

Coding Linear transformations of data, including bivariate data, can be used to make analysis of data easier. Mathematical relationships allow you to deduce values of statistical measures of the original data from the corresponding measures for the coded data.

Conditional probability When you gain new information about a situation, it may change your belief in the likelihood of the outcomes of future events. The conditional probability is the likelihood of an event after taking account of what is known about another event.

Correlation coefficient (product – moment) A measure of the linear association between two variables. It takes values between –1 and 1, and is independent of any linear change of scale of the variables.

Cumulative distribution function $F(x) = P(X \leqslant x)$ is the cumulative distribution function for the random variable X.

Discrete uniform distribution X follows a discrete uniform distribution if X takes any of the values 1, 2, 3 ..... n each with probability $\frac{1}{n}$.

Exhaustive events A set of events are exhaustive if they cover all possible outcomes.

Expectation of a random variable The mean or expected value of a random variable is defined as $\mu = E(X) = \sum px$.

Explanatory variable The independent variable, x, in the regression line of y on x.

Extrapolation Prediction for values of x which lie outside the range of values used to construct the line of regression. It needs to be used with caution.

Histogram A frequency diagram where the area of each bar is proportional to the frequency of the observations in that class interval.

Independence Two events A and B are independent if the outcome of A does not affect the outcome of B, and vice versa.

Interpolation Prediction for values of x which lie inside the range of values used to construct the line of regression. Where the line is a good fit to the data, these predictions are generally viewed as quite reliable predictions of what is likely to happen.

Interquartile range The difference between the lower and upper quartiles. It therefore represents the spread of the 'middle half' of the distribution.

Mean The mean is the sum of all the values divided by the number of values

Median The median is the middle value when the values are arranged in order

Mode The mode is the most commonly occurring value.

Model A representation of a real world situation which usually simplifies the situation so it can be analysed more easily.

Mutually exclusive events Two events A and B are mutually exclusive if they cannot occur at the same time.

Normal distribution A commonly occurring distribution in the natural world and in manufacturing processes. It is symmetric with a single peak at the centre, and often described as 'bell-shaped'.

Outliers Extreme values in a distribution. Sometimes outliers will be unimportant for the purposes of your analysis, but at other times they will be the most important values.

Prediction Substituting a value for x into the line of regression will give a predicted or estimated value for what the value of what the value of y might be. How reliable a prediction this gives will depend on whether it requires extrapolation, and on how strong the correlation is i.e. how closely the observed data fitted the regression line.

Probability A measure of belief, on a scale from 0 to 1, of the likelihood that an event will happen.

Probability distribution A set of possible values, with associated probabilities, for the outcome of a random experiment.

Probability function A formula which expresses the probability that the random variable X takes a value, k, as a function of k.

Quantiles (percentiles, deciles) The data values which divide a distribution up into any number of equal size portions. Percentiles create 100 divisions and deciles create 10 divisions etc.

Quartiles The data values which divide the distribution up into quarters. They are called the minimum, lower quartile, median, upper quartile and maximum.

Random variables (discrete) A quantity that can take any value determined by the outcome of a random event. A discrete random variable is one where the outcomes can be listed.

Regression line (least squares criteria) The straight line which provides the best fit to a set of bivariate data, using the criteria which give the minimum sum of squares of the residuals.

Relative frequency histogram A special case of a histogram where the total area is 1. This means that the proportion of data which lies between any two values will be the area between those two values.

Residual The difference between an observed y value and the value predicted by the regression line.

Response variable The dependent variable, y, in the regression line of y on x.

Scatter diagram A diagram showing the values of bivariate data as (x, y) coordinates.

Skewness A measure of the lack of symmetry of a distribution.

Standard deviation A measure of spread of a set of data. It is the square root of the variance.

Standardised score When distributions have different averages and spread, it makes it difficult to make comparisons. The standardised score takes into account the number of standard deviations above or below the mean for each data value.

Statistical modeling A representation of a real world situation which has some level of uncertainty as a central feature, and therefore needs probabilistic or statistical techniques to be used in some part of the model.

Stem and leaf diagram A way of representing a set of data which preserves the detail while showing the shape of the distribution.

Tree diagram A diagram showing the possible outcomes of two or more linked events, which helps structure the calculation of associated probabilities.

Variance The average squared distance from the mean for a set of data.

Variance of a random variable The variance of a random variable is defined as $\text{Var}(X) = E[\{X - E(X)\}^2]$

Venn diagram A diagram showing the relationship between events represented as sets, with associated probabilities.

SI

Index

SI